Space and Missile Wars: What Awaits

Space and Missile Wars: What Awaits

Nonproliferation Policy Education Center

Nonproliferation Policy Education Center
Arlington, VA 22209
www.npolicy.org

Printed in the United States of America

ISBN: 978-1-7371113-0-6

Cover image adapted from CRS-15 Mission taken on June 29, 2018 by Official SpaceX Photos on Flickr.

Nonproliferation Policy Education Center

The Nonproliferation Policy Education Center (NPEC), a 501(c)3 nonprofit organization, is a nonpartisan, educational organization founded in 1994 to promote a better understanding of strategic weapons proliferation issues. NPEC educates policymakers, journalists, and university professors about proliferation threats and possible new policies and measures to meet them.

For current publications of the Nonproliferation Policy

CONTENTS

ACKNOWLEDGMENTS

This volume of research was supported by grants from the MacArthur Foundation, the Scaife Foundation, the Carnegie Corporation of New York, and the Smith Richardson Foundation. Most, but not all, of the chapters were presented in forums NPEC cosponsored with American Bar Association's Standing Committee on Law and National Security. Brooke Buskirk is responsible for the volume's formatting and assembly. The research, of course, would not have been possible without the contributions of the volume's authors.

CHAPTER 1:

SPACE AND MISSILE WARS: WHAT AWAITS

Henry D. Sokolski

Sometimes, slow, steady changes produce revolutionary results. A case in point is missile and space technologies. Long-range missiles, satellites, and space launch vehicles used to be high technology exclusive to the superpowers. Now, scores of states have both. As for ballistic missiles and drones, even non-state actors have them, and these systems are far more accurate than anything the superpowers had even at the height of the Cold War. Then, long-range missiles could only be certain of destroying their targets if they were nuclear-armed and wiped out areas as large as cities. Now, drones are so accurate they can pinpoint and kill single individuals.

As for space satellite launchers, they originally were derived from nuclear delivery missiles. None were privately owned. Similarly, almost all space satellites were government property and, until 1965, the owners were only American and Soviet. Now, the French, Chinese, Japanese, British, Indians, Israelis, Ukrainians, Iranians, and North Koreans have all launched satellites of their own. In addition, more than 60 nations own and operate their own satellites and increasingly, satellites are launched, owned, and operated entirely by private entities.

These developments are nontrivial.

First, because of them, the ability of weak actors — smaller states and even non-state actors — to deter and harm major military powers is now greater and is likely to grow greater still. Israel, a nuclear-armed state, now views the prospect of Hezbollah upgrading its missiles' accuracy (with Iranian assistance) as a military development at least as threatening as Iran's acquisition of nuclear arms. In fact, Israel's military is so concerned, it is contemplating a massive preemptive strike against Lebanon, an attack that might well prompt Tehran to strike back at Israel. Similarly, the Houthis and other Iranian sponsored non-state actors are using ballistic missiles and drones to hit targets in Saudi Arabia and the UAE. In the case of the attack against Abqaiq on September 14, 2019, an Iranian proxy group was able to evade three layers of air and missile defenses and knockout five percent of the world's oil capacity. More debilitating attacks of this sort by weak actors are likely to come.

Second, weak actors are likely to threaten to use the limited number of accurate missiles and drones against stronger states in novel ways. They are already threatening what the Soviets called "sympathetic targets" — civilian and military assets, which if hit, will release far more energy and cause far more harm than the amount of violence required to hit them. Dams, reactors, gas and ammo depots, and petrochemical plants (e.g., Abqaiq) all would be good examples. Already, Hezbollah, the Houthis, North Korea, and China have threatened to strike reactors in Israel, the UAE, South Korea, Japan, and Taiwan. One could argue that such civilian assets have always been potential military targets. The difference now, however, is one can put these targets at risk without having to acquire and operate an advanced air force. Also, with precision, one can hit specific parts of the target (e.g., the administrative office of a power reactor that might not kill anyone) to demonstrate the seriousness of intent without necessarily causing the worst to occur. Think of precision strike as a poor man's ticket to graduated escalation.

The other novel way weak actors could use precise missiles or drones against a stronger state is to attack or threaten the assassination of the country's power elite. The most vivid portrayal of this prospect is the seven-minute video, Slaughterbots, in which facial recognition chips are integrated with small explosives on small drones and used to kill specific US senators. More prosaic demonstrations included the 2015 radiation-laced drone that landed on the roof of Japanese Prime Minister Shinzo Abe's office and the 2018 drone assault of Venezuelan President Nicolas Maduro. These sorts of attacks are likely to become more attractive and frequent. With any bad luck, weak actors may be able to "kill" a nation, not by destroying its military or its key cities or industries, but by merely taking out a few scores or hundreds of key officials with precision facial recognition-assisted drones.

This, then, brings us to yet another way the proliferation of accurate missiles and drones may change the face of war — the increased likelihood of more Sarajevos, i.e., of catalytic wars. In 1914, most military analysts knew who the major combatants of any future war in Europe might be — the UK, Germany, France, Russia, and the Austria-Hungarian Empire. What nobody quite anticipated, however, is how the war might begin (in this case, with an assassin's bullet fired at an archduke in Serbia). Our future may rhyme.

With the proliferation of accurate strike systems to an ever-longer list of weak actors, count on there being ever more opportunities for Sarajevo moments. Hezbollah missiles strike Haifa. Israel strikes missile attack staging areas in Lebanon. Iran responds by striking select Israeli military targets. Israel, then responds by striking targets in Iran. Russia sides with Iran. The United States sides with Israel. At this point, one could easily imagine a replay of the Suez crisis in which two superpowers (the US and Russia) face each other in a nuclear standoff, this time in the Gulf. Consider this example and reflect on India and Pakistan (and the myriad of discontented proxies each could arm) or what North Korea might do with precise conventional missiles against civilian targets in South Korea or Japan. One might add the greater Middle East, Central Asia, the Balkans, and the Baltic. As precision strike systems spread, so too will the list of possible catalytic and potentially catastrophic war scenarios.[1]

---

1. See Henry Sokolski, "Dr. Strangelove's New Passion: Precision-Guided Mayhem," American Purpose, March 17, 2021, at https://www.americanpurpose.com/articles/dr-strangeloves-new-passion-precision-guided-mayhem/.

Of course, precise missile and drone attacks are only as good as their targeting information, which increasingly relies on sensors and cameras based in space. Thus, all of the major militaries of the world have launched satellite systems of their own. Just as important, the world's civil and commercial sectors are now highly dependent upon space-based imagery (e.g. Planet), navigation systems (e.g., GPS), and communications relay systems. In the near future, internet and communications relays consisting of thousands of small satellites in low-Earth orbit will drive economic growth. Once deployed, these commercial systems and their private space launch systems will be used more and more for military purposes.

Similarly, civilian space activities, once kept relatively separate from the military are being conflated with it. Civil lunar exploration and related activities can and will be exploited to hold military satellites at geosynchronous and lower orbits at risk. These satellites include some of the world's (and America's) most important military systems — navigational; command, control, and communications; and early warning satellites.

Compounding this conflation is the increasing number of dual-use capable space systems. These include: ground-based lasers used for ranging and other civil purposes that can also be used to dazzle, blind, or damage satellite systems; and rendezvous satellites that can refuel, repair, and reposition satellites or de-fuel, damage, and take them out of position. Combine each of these developments with the increasing number of states and private entities that are engaged in space launch and satellite ownership and operation and you set the stage for space "conjunctions" and interactions both welcome and hostile.

With so many new space actors, attribution of any unwanted action in space will be difficult. Nor will it help that there are few, clear agreed redlines for international space behavior. The major powers have generally favored maintaining secrecy and military options in space over clarifying what clearly constitutes illicit, hostile actions. Again, this too is a prescription for strategic miscalculation. As the front lines of strategic deterrence gravitate toward the satellite systems that the world's navies, air forces, armies, and economies rely on, space will become the first field of battle. Clarifying what competition and conflict in space may look like is difficult.

This volume of essays is designed to help. It is divided into two sections. The first focuses on space; the second on missiles. Peter Garretson's first chapter projects what war in space might look like one to two decades from now. Brian Chow's second chapter considers how private space firms' desire to avoid satellite collisions (and the loss of their satellites) might encourage adherence to space rules of the road that the world's civil, commercial, and military space sectors might follow. Xavier Pasco's third chapter details French space policies which include the world's clearest articulation of why space keep-out zones will be necessary and how they might best be enforced. The last space-focused chapter is the winning US collegiate national debate submission on space warning zones as a diplomatic approach to dealing with the rendezvous satellite threat.

The balance of the volume is focused on missiles and drones, how they might be used and controlled. The first of these chapters, mine, is on the future of missile and drone warfare. The second, David Cooper's, is on why missile controls are both needed and unlikely. Finally, the third, John Maurer's, is a guide to how one might coordinate increased missile competitions with controls.

More, of course, could and will be written. This volume, however, is a start.

Chapter 2:

What War in Space Might Look Like Circa 2030-2040

Peter Garretson[2]

Introduction

This essay seeks to give the reader two distinct visions of space warfare circa 2030-2040. First, a conservative view presumes wars begin on Earth (over geo-strategic interests on Earth) and extend into space to maintain the advantages of continued overhead satellite sensing. The second view is expansionist. It presumes space wars begin in space over geo-strategic interests in space and proceed with relative independence of events and military activity on Earth. Actual conflict of course might very well see a mix of both, but they are presented as extremes, or ideal-type cases to help the reader to see the outlines of different causal logics driving each. At some point, circa 2030 (and perhaps even sooner), economic development activities in deep space will create a truly unprecedented situation: nation-states will have geo-strategic interests on other planetary bodies — the moon and asteroids. The degree to which the future resembles the expansionist view will depend on how much and at what speed activities are conducted in deep space and on the Moon in particular. Those who would dismiss the expansionist scenarios as 'too far out' should consider that already today a number of states are talking about Lunar mining, and that the United States,

2. DISCLAIMER: This is a speculative work by the author. The reader should not construe these speculations to represent positions or policy of the United States. While the author spent part of his career in the US Department of Defense, his views diverge significantly from current DoD views on these matters.

China, India, Russia, Japan, and Israel are already putting precursor missions on the Moon. A wholistic view of both has implications for military doctrine, concept design, force structure design, and international governance including arms control and the law of armed conflict, as well as for legislation and policy. This suggests that 1) military doctrine must anticipate that United States geo-strategic interests will likely encompass economic activities in deep space, and develop concepts for peacetime strategic offensives to maintain positions of advantage; 2) that design of a 2030-2040 force structure requires attention to the unique navigational, maneuver, logistics, and power projection needs for deep space vehicles; 3) consideration and anticipation of conflict may enable the community of nations to arrive at consensus on certain conflicts or types of conflict they mutually wish to avoid, and 4) a role for civilian leadership to ensure the US DoD is prepared through legislation and policy to specify roles and missions to protect space commerce, to specify plans, and to specify the Area of Responsibility (AOR) as encompassing the entirety of the Cis-Lunar theater.

Below, I provide scenarios for crisis and conflict. The conservative view sees space war taking place close to Earth, within the Earth's Gravity well in Low Earth Orbit (LEO) or the Geostationary belt (GEO). In contrast, the expansionist view considers conflict that might occur in proximity to Earth's Moon, *approximately ten times as distant as GEO (See figure 2.01 below)*. Readers curious how I arrived at these can consult the appendix.

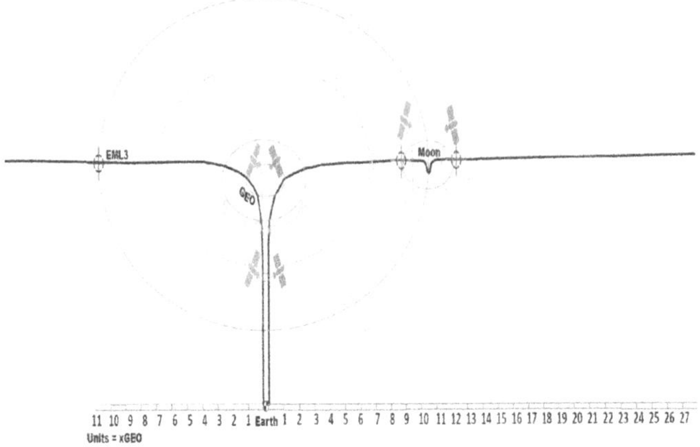

Figure 2.01. Two Theaters of Space Crisis & Conflict (Source: Author).

Figure 2.01 provides a basic 'terrain' map to orient the reader to the two theaters in which space conflict circa 2030-2040 might take place. The key features of this map are to-scale depictions of distance and the relative strength of gravity of the Earth and the Moon – depicted as gravity wells. Near the Earth, the Earth's gravity dominates spacecraft trajectory, labeled as *Earth Dominant Space.* Near the Moon the Moon's gravity dominates, labeled as *Moon Dominant Space.* In between is an area of shared gravitational influence, where the Moon can perturb spacecraft, and the *Earth-Moon Influence Boundary* above which the gravitational influence of the Moon is must be taken into account. Also labelled are the Earth-Moon Lagrange Points (EML1, EML2, EML3, EML4, and EML5) where gravitational and centrifugal forces balance such that spacecraft may be 'parked' at these points and maintain their relative position to the Earth and Moon, and may constitute strategic locations for commerce or control. The conservative scenarios take place entirely within *Earth Dominant Space*, and in fact GEO and below. In contrast the expansionist scenarios take place beyond GEO, and mostly in *Moon Dominant Space.*

A Conservative View of Future Space Warfare: War Extends to Space

The conservative view of space warfare is that 'there is no such thing as a space war, there is just war'[3] and that a terrestrial 'war extends into space.'[4] Here the assumption is that the primary value of space assets is to support the terrestrial warfighter, and space control is a necessary activity to secure essential space information services (weather, missile warning, position, navigation, timing, blue force tracking, overhead surveillance, beyond line-of-sight communications). Deterrence by punishment is assumed to be ineffective[5] since the US has more to lose and the opponent can operate effectively in their near abroad without space. Because all military satellites are now held at risk, satellite protection and deterrence by denial is the key strategy. Therefore, the evolution of the architecture is toward a proliferated constellation of Low Earth Orbit (LEO) satellites, and better attribution and perhaps local bodyguards in GEO. The potential threats are manifold, with ground-based anti-satellite missiles and in-space robotic 'snugglers'[6] likely to multiply. Space-based terrestrial strike weapons[7] and through-space transportation

---

3. Steve Hirsch, "There is no 'War in Space'," *Air Force Magazine*, May 29, 2018, accessed June 30, 2020, https://www.airforcemag.com/article/there-is-no-war-in-space/.
4. Marcia Smith, "U.S. Space Command Reestablished After 17-Year Hiatus," *Space Policy Online*, August 29, 2019, accessed June 30, 2020, https://spacepolicyonline.com/news/u-s-space-command-reestablished-after-17-year-hiatus/.
5. John Klein, "Towards a Better U.S. Space Strategy: Addressing the Strategy Mismatch," *Strategy Bridge*, September 9, 2019 Accessed June 30, 2020, https://thestrategybridge.org/the-bridge/2019/9/9/towards-a-better-us-space-strategy-addressing-the-strategy-mismatch.
6. Brian G. Chow, "Stalkers in Space: Defeating the Threat," *Strategic Studies Quarterly*, Summer 2017, accessed June 30, 2020, https://www.airuniversity.af.edu/Portals/10/SSQ/documents/Volume-11_Issue-2/Chow.pdf.
7. Staff Writers, "US Project Thor would fire tungsten poles at targets from outer space," *Space Daily*, November 12, 2018, accessed June 30, 2020, https://www.spacedaily.com/reports/US_Project_Thor_would_fire_tungsten_poles_at_targets_from_outer_space_999.html ; Blake Stilwell, "These Air Force 'rods from god' could hit with the force of a nuclear weapon," *We are Mighty*, May 22, 2020, accessed June 30, 2020, https://www.wearethemighty.com/articles/these-air-force-rods-from-god-could-hit-with-the-force-of-a-nuclear-weapon.

and insertion remain of interest, but likely will not compete well against other power-projection mechanisms. Evolution of the order of battle is entirely military, though it likely leverages commercial launch and very large low-Earth Orbit (VLLEO) mega-constellation capabilities.[8] The architecture is entirely 'geocentric.' Deep space is conceptualized as the domain of NASA. Capabilities which we might expect to see in this time frame might include space-sensor layers to detect and track hypersonics,[9] and keep custody of other moving objects.

Conservative Considerations in Crisis and Coercion

Several considerations affect the lead-up and logic of how and why a crisis may escalate and embroil the participants in a difficult to exit spiral.

Escalation Route: The primary route to escalation is that an adversarial nation would take action with regard to a US partner's equities which compromise the US standing as a reliable partner or directly call into question US treaty defense obligations. The US would then posture to show the aggression is unacceptable.

Dominant Strategy: The adversary would then have to consider both crisis management and bargaining, the potential for a regional limited war, and escalation to general war where it was uncertain of US restrictions on regime change. The dominant strategy in all cases

8. Casey Handmer, "The SpaceX Starship is a very big deal," *Casey Handmer Blog*, October 29, 2019, accessed June 28, 2020, https://casey-handmer.wordpress.com/2019/10/29/the-spacex-starship-is-a-very-big-deal/#:~:text=Starship%20is%20not%20just%20a,This%20is%20a%20big%20deal. ; Casey Handmer, "Starlink is a very big deal," *Casey Handmer Blog*, November 2, 2019, accessed June 28, 2020, https://caseyhandmer.wordpress.com/2019/11/02/starlink-is-a-very-big-deal/.

9. Peter Garretson, "U.S needs a space sensor layer to protect against hypersonic missiles," *Washington Times*, June 22, 2020, accessed June 30, 2020, https://www.washingtontimes.com/news/2020/jun/22/us-needs-a-space-sensor-layer-to-protect-against-h/.

for a US adversary is to remove the force-multiplying effects of US space infrastructure.

Raising the Costs of Intervention: The key action in a crisis is to raise the costs for the US globally, and to signal its credible capability to hold at risk systems which might be perceived as more valuable than the US interests in counter-intervention. We could expect then a series of postures that remind the US of its vulnerability. Co-orbital platforms would snuggle in,[10] ground-based ASAT missiles would show increased readiness and mobility, laser systems would illuminate targets, denial of service of symbolically important channels via jammers[11] or cyber-attack,[12] and tests of new systems might be carried out. We might also expect an "accidental collision" or even a "friendly inspection"[13] of a manned US government or commercial space station. All to say, "we can touch you...back down." Some of this is likely to be done in ways not visible to the public, but in the US system such private signaling is still likely to leak out. The US could not be seen as publicly backing down from such threats, so it would have to counter-threaten in public while privately offer-

10. Brian G. Chow Space, "Nuclear vulnerability: In-orbit bodyguards would help protect NC3 satellites from attacks," *Space News*, April 1, 2019, June 30, 2020, https://spacenews.com/op-ed-nuclear-vulnerability-in-orbit-bodyguards-would-help-protect-nc3-satellites-from-attacks/.

11. Trevor English, "U.S. Space Force's First Offensive Weapon Is a Satellite Jammer: The Space Force is ready for battle," *Interesting Engineering*, April 22, 2020, accessed June 30, 2020, https://interestingengineering.com/us-space-forces-first-offensive-weapon-is-a-satellite-jammer#:~:text=The%20new%20weapons%20system%20delivered,in%20a%20matter%20of%20minutes.

12. Zak Doffman, "U.S. Military Satellites Likely Cyber Attacked By China Or Russia Or Both: Report," *Forbes*, July 5, 2019, accessed June 30, 2020, https://www.forbes.com/sites/zakdoffman/2019/07/05/u-s-military-satellites-likely-cyber-attacked-by-china-or-russia-or-both-report/#4679ea02dd32 ; William Akoto, "Hackers could shut down satellites -- or turn them into weapons," *GCN*, February 12, 2020, July 30, 2020, https://gcn.com/articles/2020/02/12/hackers-satellites.aspx.

13. By Chelsea Gohd, "2 Russian satellites are stalking a US spysat in orbit. The Space Force is watching," *Space.com*, February 11, 2020, accessed June 30, 2020, https://www.space.com/russian-spacecraft-stalking-us-spy-satellite-space-force.html

ing a de-escalatory program.[14] The US is likely to take comparable measures, as well as to conduct horizontal escalation in an adjacent domain.

Fog and Uncertainty: The situation is likely to degrade further, as one or both sides attempt to mask their signatures through release of decoys, dazzling and jamming, and maneuver to change their orbital elements.

Conservative Scenarios for Space Warfare

Incentives to Escalate & Pre-Empt: The problem with the adversary's threats to US space systems is that the US must take them seriously, and realize as well that an actual strike on US space systems would so materially change conditions that it might feel it must pre-empt. It is not an exaggeration to think that a significant degradation of space could render the US military a tenth to 100th as effective by which we mean that the US would need to deploy 10 to 100x the level of activity to accomplish the same effect. For example, in WWII, it took an average of 100 sorties to destroy a single target while today one sortie can destroy four or five.[15] Numerous specific degradations are foreseeable. A loss of missile warning would make airbases and ships more vulnerable to missiles, enabling a potentially decimating first strike against forward-deployed (and perhaps even CONUS based) forces. A loss of timing signal[16] would remove the ability to reliably signal-hop or have secure communications. A loss of GPS would severely reduce the accuracy of US missiles. A loss of beyond line-of-sight satellite communications would impede

14. As was the case with Cuban Missile Crisis.
15. Martha Harbison, Emily Elert, Clay Dillow, "Infographic: Horses And Bayonets? Guns And Butter? Is our Navy really the smallest it's been since 1916? And what does that mean, anyway?," *Popular Science*, October 23, 2012, accessed June 28, 2020, https://www.popsci.com/technology/article/2012-10/horses-and-bayonets-guns-and-butter/.
16. "Timing," GPS.GOV, n.d., accessed June 28, 2020, https://www.gps.gov/applications/timing/#:~:text=GPS%20receivers%20decode%20these%20signals,economic%20activities%20around%20the%20world..

any coordination or update between dispersed forces or allow update of missiles. A loss of overhead sensing would severely reduce US knowledge of enemy order of battle, and virtually eliminate the ability to prosecute mobile targets. Finally, most worrisome, the US would have to take seriously that a loss of nuclear command and control or missile warning could facilitate or signal a surprise nuclear first strike and delaying retaliation sufficiently to further attrite US nuclear forces.

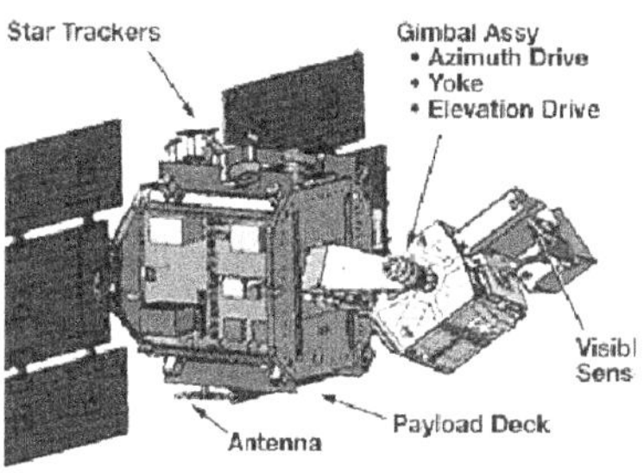

Figure 2.02. SBSS-1 spacecraft and its SBV imager (image credit: Boeing Company).[17]

What are Attractive Targets?: Targeting of space systems relies on a belligerent's space domain awareness (SDA),[18] command and control systems, and anti-satellite weapons systems (an example of one SDA system is depicted in figure 2.02 above). Military planners

17. "Space Based Space Surveillance," Mission Brochure, Boeing Company, URL: http://www.boeing.com/defense-space/space/satellite/MissionBook.pdf cited in: "SBSS (Space-Based Surveillance System)," EOPortal, n.d., accessed July 25, 2020, https://eoportal.org/web/eoportal/satellite-missions/content/-/article/sbss#foot37%29.
18. For an in-depth discussion of Cis-Lunar SDA, see Sean McClain, "Celestial Sentinels: A Framework For Cis-Lunar Space Domain Awareness In 2035," (masters thesis), Air University, March 2020, accessed June 30, 2020, https://www.linkedin.com/posts/namratagoswami_celestial-sentinels-by-sean-mcclain-activity-6664614183398629376-K-IH.

are likely to favor targets which provide the maximum effect, enable the greatest economy of force, can be struck with the highest likelihood of success, and generate the minimum collateral damage and inadvertent escalation. As both nations multiply their sensors, links, command posts, and weapons, what appears most attractive and most vulnerable can shift. In the conservative scenario, the most critical and central nodes are on the ground, and can be struck only by terrestrial forces: clandestine or special operations forces, naval forces, or air forces.

Moreover, the belligerent forces of both sides are threatened by the adversary's space-enabled reconnaissance strike system. For example, US major power-projection legacy assets such as a carrier or combat aircraft are moving targets. Targeting them is greatly facilitated by overhead intelligence, whether optical, infra-red, radar, or RF mapping.[19]

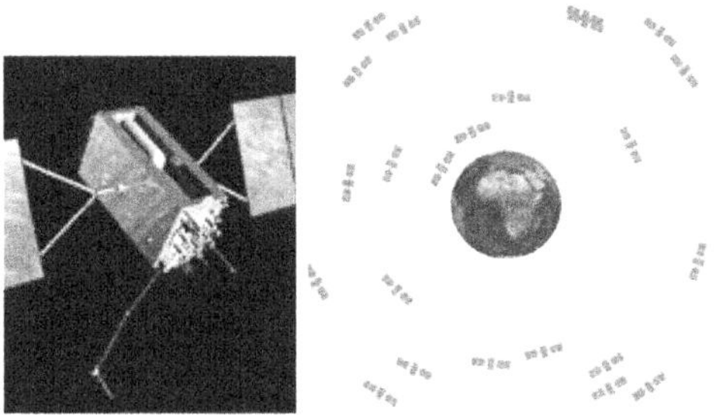

Figure 2.03 & 2.04. An example of a global navigation satellite system (GNSS) Satellite (GPS Block III) and a full GNSS constellation of 24+ GPS satellites (source GPS.gov).[20]

Once observed, attacking platforms must be passed targets and updates via satellite communication, and must guide themselves where a global navigation satellite system (GNSS[21]) is a great aid. For these

19. Of note, all of these have some level of commercial availability.
20. "Space Segment," GPS.gov, n.d., accessed August 11, 2020, https://www.gps.gov/systems/gps/space/#IIIF.
21. Gps.gov, Other Global Navigation Satellite Systems (GNSS), n.d., accessed

reasons, both sides have an incentive to target the enemy satellite system, for which it requires SDA,[22] and for which it must degrade adversary SDA to reduce their warning time and ability to take counter measures. Counter-space forces will be under severe pressure from terrestrial warfighters (naval, land forces, and air forces) to reduce their vulnerability by destroying space-based surveillance & reconnaissance systems.

Economy of Force Criteria: However, both sides are likely to face a scarcity of resources,[23] preventing all desired military targets from being simultaneously struck. Therefore, the logic of the situation dictates that you offensively go after the systems which would most reduce your own capability, and defend those systems which maximize your long-term ability to cumulatively prosecute enemy forces. If the opponent has multiple, resilient means of securing back-up services, you may leave those systems untouched. For example, the future ubiquity of target-quality overhead optical sensing, the ubiquity of global navigation systems, and perhaps even satellite internet may mean that attacking sovereign assets of these sorts are just wasted bullets in the initial volley.

Illustrated in extremely simple terms, if an opponent's reconnaissance strike complex for an anti-satellite missile consisted only of 1 ground-based sensor, one command center, one data-link, and 1 anti-satellite missile, striking any one of them would deny the op-

---

June 30, 2020, https://www.gps.gov/systems/gnss/.

22. The US Space Force defines SDA as the "identification, characterization and understanding of any factor, passive or active, associated with the space domain that could affect space operations and thereby impact the security, safety, economy or environment of our nation" per Sandra Erwin, "Air Force: SSA is no more; it's 'Space Domain Awareness'," *Space News*, November 14, 2019, accessed June 30, 2020, https://spacenews.com/air-force-ssa-is-no-more-its-space-domain-awareness/.

23. It is almost always the case that you have more targets than you have munitions, though it is also common for states to underfund munitions: Aaron Mehta, "The US is running out of bombs — and it may soon struggle to make more," *Defense News*, May 22, 2018, June 30, 2020, https://www.defensenews.com/pentagon/2018/05/22/the-us-is-running-out-of-bombs-and-it-may-soon-struggle-to-make-more/.

ponent the ability to strike. If all of the targets could be attacked and destroyed with a bomb, and the attacking state had just one bomb, one might choose the target hardest to reconstitute. If on the other hand, the opponent had 1 ground-based SDA sensor, 2 command centers, 3 data-links, and 10 anti-satellite missiles, the most economical way to deny the system is to destroy the single point of failure, the SDA sensor. In reality, command posts are often very hard targets (hardened and deeply buried, deep within a belligerent's homeland), ASAT weapons are expensive (but may be mobile, proliferated and difficult to track), uplink/downlink/crosslink datalinks can be proliferated at low cost; in contrast, Earth-based sensors are often fixed and visible, and space-based sensors are likely visible and jammable.

Counter-Force, Counter-Value, and Astroeconomic 'Guerre de Course': Attacks on space systems may not be limited to military-only systems. As in nuclear thinking, we can make a distinction between counter-force attacks which target military systems to reduce military capability, and counter-value attacks which directly injure the broader society.[24] As a form of strategic attack or coercion, a belligerent could choose some level of guerre de course or astroeconomic warfare,[25] which injure the other belligerent or impose third party (allied or neutral) costs. Some military targets like GPS are also critical civilian and commercial infrastructure and would constitute astroeconomic targets.

Four Possible Outcomes: Four general outcomes appear possible: mutual denial, relative invulnerability, space supremacy, and space superiority. It is possible that both sides might have such terrestrial counter-space capabilities that neither side can enjoy the benefits of space, and the advantages accrue to the side whose force structure is better positioned to express military power in the relevant theatre without space information services. It is theoretically possible

---

24. See Encyclopedia Britannica, "Counterforce doctrine: nuclear strategy," n.d., accessed June 29, 2020, https://www.britannica.com/topic/counterforce-doctrine.
25. This phrase appears to have been coined by Bleddyn E. Bowen, *War in Space: Strategy, Spacepower and Geopolitics*, (Edinburgh: Edinburgh University Press, 2020), p. 124.

(though very unlikely given current trends) that both sides could attrite each other's ability to target satellite systems, or that satellite defenses were far more capable than attacking systems, leaving space a relative sanctuary.[26] It is possible that one side could succeed in destroying or denying an adversary's space information services and access to backups, and succeed in destroying the adversary's counterspace reconnaissance strike system to such an extent that it is possible to operate in space with impunity, in which case it would have space supremacy. The final possibility is a more limited and temporal degree of space control,[27] where there is sufficient advantage to enable sufficient survival and freedom of action to employ space information services not everywhere but locally for specific objectives in time and space.

Top Priorities in Space War: If space services can be secured for one side, then it allows full exploitation of those services. To attack an enemy's space-based reconnaissance strike complex,[28] you require your own SDA, and must deny them theirs so they cannot blunt your attack. Loss of SDA is the loss of one's eyes to see and target, and so its denial effectively denies all subsequent precision attack. Therefore, we can expect the primary target of both belligerents to be the SDA of the other. For belligerents actively engaged in terrestrial combat, arguing for equal priority will be the denial of sensors and data-links which enable precise targeting, especially of mobile targets. To the extent one's own offensive forces require space information services to function with precision, their sensors and datalinks are likely to be defended assets. To the extent the security of one's terrestrial forces depends heavily on advance warning of a strike (for example warning of incoming missiles or warn-

26. Brian Weeden, "The End of Sanctuary in Space," *War is Boring*, January 7, 2015, accessed June 29, 2020, https://medium.com/war-is-boring/the-end-of-sanctuary-in-space-2d58fba741a.
27. See official definition and discussion of Space Control: Department of Defense, *Joint Publication 3-14 Space Operations*, April 10, 2018, accessed June 30, 2020, https://www.jcs.mil/Portals/36/Documents/Doctrine/pubs/jp3_14.pdf.
28. Barry Watts, *The Maturing Revolution in Military Affairs*, June 2, 2011, accessed June 30, 2020, https://csbaonline.org/uploads/documents/2011.06.02-Maturing-Revolution-In-Military-Affairs1.pdf.

ing of incoming bombers), its sensors and datalinks are likely to be defended assets. Freedom of action for each side requires defense of their offense-enabling systems, and attack of adversary defense-enabling systems and their local defenses.

Technological Trends Which May Alter These Scenarios: Over time, the capability and relative cost and economies of anti-satellite weapons may shift. Overall, such shifts are unlikely to materially alter the overall logic of the situation except in one specific instance: the introduction space-based long-range direct fire anti-satellite weapons. The introduction of space-based, counter-space direct-fire weapons (such as lasers, neutral particle beams, or rail guns) could alter the situation such that these are the major impediment to freedom of action, and are fewer in number than the sensors that direct them, being both the most dangerous and most economical part of the system to attack. In such a case, they would become the primary first target to enable follow-on freedom of action, not unlike the contest between fighter aircraft for air superiority, or between battleships in WW1 and later carriers in WW2 for command of the sea.

How the War Unfolds

The war thus unfolds in one of four obvious ways.

Scenario 1 'He's too close' [in space]: At some point in crisis posturing, a belligerent gets too close to a capital asset in space, and the use-it-or-lose-it pressure becomes unbearable. It triggers self-defense and simultaneously broad-based offensive counter-space effort to prevent adversary counter-space activity.

Scenario 2 'Warning Shot': As part of the posturing, a warning shot or limited attack is fired to prove seriousness of intent. The receiver is likely to perceive this instead as a signal that this is the first indication of a more general war. Alternately, unsure that this is just a warning shot and not an actual attack, the receiver may attempt to ensure the safety of the targeted asset by interrupting the 'kill chain' by attacking supporting satellite infrastructure, whether reconnaissance, data-link, or navigation satellites. As part of such self-defense measures, it may also degrade SDA or attack in-space defending assets. Such actions would be perceived by the other side as an actual attack and escalation, triggering their own broad based counter-space plan.

Scenario 3 'First Strike': As intervention forces (probably the US) flow forces into the region to posture, there is a point that such a posture becomes too threatening. Once the decision is made that to win, one must attack first with an overwhelming strike to kill the logistics and basing to shorten the legs of the intervening forces, then one must maximize surprise and ensure the space information services for precision strikes. The first move then is to destroy or suppress missile warning an SDA so the intervening force cannot see or alert its forces, nor see what has taken place in space, nor reliably attack the supporting reconnaissance and precision guidance.

Scenario 4 'Pre-empting First Strike': Fearing, and seeing indications of scenario 3, and knowing that allowing the other side the first offensive strike would be game-winning, the intervener decides to pre-empt, and attack the enemy's SDA and space information services.

Of note, in all of the scenarios, we would expect very early attack of SDA and attack on space-based sensors and guidance to enable prosecution of mobile targets. Because SDA sensors are both on the ground and in-space, success requires a multi-domain coordinated counter-SDA campaign.

Terrestrial Components of the Counter-Space Campaign

In such scenarios, the bulk of the 'space war' is taking place on planet Earth, and most of the attacks on spacecraft are launched from planet Earth.

Concern that one might not be able to fully destroy the SDA sensors, and the difficulty of attacking hardened command facilities or proliferated links makes attacking ground-based ASAT facilities themselves an additional high priority. For example, attacking all the sensors might mean you have disabled all the missiles, but any surviving sensors might enable all missiles, whereas striking an ASAT facility means that that facility means one less effector to complete the kill. Belligerents must assume that any fixed site or known-location ASAT facilities (ground based lasers,[29] ground based missile kinetic ASATS[30]) would also be targeted, so there is an incentive to 'go big early' and use such capabilities to their maximum until they are used up or lost to adversary action. Despite the likely location of such systems deep in a belligerent's homeland, the assumed 'deep magazine' of laser threats, and their ability to rapidly and continually attrite LEO satellites as they fly overhead could create extreme incentives to strike them very early in a conflict.

Denying Third Party Space Information Services: The existence of 3rd-party and commercial SDA services to the opponent likely would also be unacceptable. Extremely timely and strong incentives would likely result in cyber-based 'blockades' of such services contributing to enemy action. Similar 'wars' for commercial access would play out across the range of space information services, resulting in cyber and electronic-warfare-based denial of services (even

29. Bill Gertz "China may deploy anti-satellite laser weapons next year able to destroy U.S. military satellites," *Washington Free Beacon*, February 18, 2019, https://www.militaryaerospace.com/trusted-computing/article/16711585/china-may-deploy-antisatellite-laser-weapons-next-year-able-to-destroy-us-military-satellites "Russia Lasers," *Global Security*, n.d., accessed July 1, 2020, https://www.globalsecurity.org/space/world/russia/lasers.htm; Bart Hendrickx, "Peresvet: a Russian mobile laser system to dazzle enemy satellites," *The Space Review*, June 15, 2020, accessed July 1, 2020, https://www.thespacereview.com/article/3967/1.
30. U.S. Space Command Public Affairs, "Russia tests direct-ascent anti-satellite missile," April 15, 2020, accessed July 1, 2020, https://www.spacecom.mil/MEDIA/NEWS-ARTICLES/Article/2151611/russia-tests-direct-ascent-anti-satellite-missile/; Doug Messier, "DIA: China Has Ground-based ASAT Missile, Likely Working on Laser Weapon," *Parabolic Arc*, February 18, 2019, accessed July 1, 2020, http://www.parabolicarc.com/2019/02/18/dia-china-groundbased-asat-missile-working-laser-weapon/ ; Jeff Foust, "India tests anti-satellite weapon," *Space News*, March 27, 2019, accessed July 1, 2020, https://spacenews.com/india-tests-anti-satellite-weapon/.

to neutrals[31]) and involve financial and legal coercion and strong political pressure if time allowed, and buying up services wherever possible. Insider threats and sabotage should also be expected.

Preventing Reconstitution: Once the decision has been made to deny space-based services, it could be unacceptable to allow the other side to reconstitute during the conflict. Therefore, both sides would have incentives to deny the other launch sites and launch opportunities through suppression, sabotage, or strike.

Cross-Domain Integration: What ought to be clear from this discussion is the high level of integration required by the supporting terrestrial forces (clandestine, SOF, air forces, naval) early in the campaign, and the high degree of coordinated war-planning required to take a large diversity of counter-space actions quickly and in the right sequence. Without the equivalent of Strategic Air Command's Single Integrated Operational Plan (SIOP)[32] to attempt space supremacy, the ensuing fumbling is likely to leave the unprepared side at a significant disadvantage. This may prove difficult, and space services are relatively new, and may lack the necessary political clout to argue for priority, and establish themselves as the supported effort in the early phases of conflict.

Prompt Global Strike (PGS) and Spaceplanes: Since the Cold War militaries have considered the possibility of delivering conventional munitions from or suborbitally through space.[33] Like hypersonic weapons generally, the advantages are: range, speed of transit,

---

31. For an excellent discussion of the likely violation of neutrals in space war, see Bleddyn E. Bowen, *War in Space: Strategy, Spacepower and Geopolitics*.
32. See Encyclopedia Britannica, "Single Integrated Operational Plan: United States warfighting plan," n.d., accessed June 29, 2020, https://www.britannica.com/topic/Single-Integrated-Operational-Plan.
33. Staff Writers, "US Project Thor would fire tungsten poles at targets from outer space," *Space Daily*, November 12, 2018, accessed June 30, 2020, https://www.spacedaily.com/reports/US_Project_Thor_would_fire_tungsten_poles_at_targets_from_outer_space_999.html; Blake Stilwell, "These Air Force 'rods from god' could hit with the force of a nuclear weapon," *We are Mighty*, May 22, 2020, accessed June 30, 2020, https://www.wearethemighty.com/articles/these-air-force-rods-from-god-could-hit-with-the-force-of-a-nuclear-weapon.

and kinetic energy on arrival without nuclear fall-out. The low response time, difficulty to defend, and low collateral damage makes them attractive conventional first strike weapons, especially against timely and hard targets. Because of the high level of damage that an opponent's SDA-ASAT system can inflict, literally every minute may count, and the incentive is to attack it with the fastest system available. An obvious first use then is to use these to attack the enemy's SDA system and capable systems (such as ground-based lasers) on your first volley. However, if the enemy also has such weapons, you likely need to strike them first to allow further freedom of action, or they are going to strike yours. This creates a destabilizing condition analogous to ICBMs or IRBMs where each side if vulnerable, where offense-dominance incentivizes first strike, and where warning times are short.

Restraint & Unrestraint

How exactly belligerents choose to attack each other's counter-space reconnaissance strike complex depends on many factors.

Incentives to Global War: A key problem is that space constellations are not confined to a single theater but serve global interests. Therefore, an attack on space systems creates a very high probability that the war will be seen, at least by the US, as a general war affecting the totality of its global equities. Overall, the offense dominance of current systems creates an incentive for pre-emption on both sides.

Incentives Toward Restraint: The general undesirability of a general war may tempt actors toward self-restraint, as was seen in the politically constrained use of airpower in Korea, Vietnam, and India-China and India-Pakistan conflicts. At its most extreme, we could imagine a head-of-state to head-of-state phone call agreeing that space was off-limits to anything but ground-based jamming in the theater. We could also see extremely carefully choreographed movements akin to the Cuban missile crisis.[34] Escalation concerns (such as unwanted confusion regarding nuclear retaliation) could place certain Nuclear Command and Control (NC3) and missile warning satellites on the no-strike list[35]—but this cannot be taken for granted[36] if these systems support tactical users and if one side is willing to gamble the other side knows it is a limited war. Escalation concerns could make terrestrial facilities in the adversary homeland or third party off-limits to kinetic strikes.

Dialectic Response

Even in a limited exchange involving kinetic antisatellite weapons, such as two sides each demonstrating they can attack the other, an enormous amount of space debris will be created. It is unlikely that humanity would abandon important orbital regions. Whatever might be the effect on arms control or the law of armed conflict weapons legality,[37] it is likely there would be a significant acceleration[38] in

---

34. Editors, "Cuban Missile Crisis," *History.com*, January 4, 2010, accessed June 29, 2020, https://www.history.com/topics/cold-war/cuban-missile-crisis.
35. In fact some, authors think the US should deliberately disaggregate its NC3 from other systems: From Sanctuary to Battlefield: Elbridge Colby, *A Framework for a U.S. Defense and Deterrence Strategy for Space*, CNAS January 27, 2016, accessed June 30, 2020, https://s3.amazonaws.com/files.cnas.org/documents/CNAS-Space-Report_16107.pdf?mtime=20160906081938.
36. Brian G. Chow, "Nuclear vulnerability: In-orbit bodyguards would help protect NC3 satellites from attacks," *Space News*, April 1, 2019, accessed June 30, 2020, https://spacenews.com/op-ed-nuclear-vulnerability-in-orbit-bodyguards-would-help-protect-nc3-satellites-from-attacks/.
37. David Stuckenberg and Peter Garretson, "Updating the Laws of War for the 21st Century: Time to Reconsider, and Ban, Modern Nation-Killing Acts," The Mitchell Institute, April 2, 2019, accessed July 1, 2020, https://www.mitchellaerospacepower.org/single-post/2019/04/02/Updating-the-Laws-of-War-for-the-21st-Century-Time-to-Reconsider-and-Ban-Modern-Nation-Killing-Acts.
38. In the wake of a significant increase in debris, incentives might exist to create a new regime: Peter Garretson, Alfred B. Anzaldúa, and Hoyt Davidson, "Catalyzing space debris removal, salvage, and use Maritime lessons and a proposal for a space salvage entity and pollution credit system," *The Space Review*, December 9, 2019, accessed July 1, 2020, https://www.thespacereview.com/

Active Debris Removal (ADR).[39] This has two perverse effects: first, that the future is more 'safe for kinetic ASATS' and second, that a future conflict begins with a broader proliferation of new counter-space capabilities both robotic and directed energy.

Any contingency involving destruction of in-space infrastructure is likely to resolve many uncertainties and incentivize diverse actions. Even the winner of space supremacy is likely to find that victory is ephemeral without a major series of new investments. The victor may be left with a capable on-orbit system while the loser may have none. But if the loser retains any meaningful economy and industry, they can rapidly set to work on a first-class space denial capability custom built against the now aging and static constellation of the winner. Unless the winner is willing to take consistent pre-emptive action risking war, they will not be able to stop the proliferation of counterspace capabilities or the reconstitution of orbital capabilities. To do so, they would need to have the will and capability to go 'full Dolman'[40]—to orbit both a space-based launch-interdiction capability, and a space-based strike capability to rapidly suppress any counterspace capabilities. Otherwise, over time, the losing nation could create a posture to deny space from planet Earth.

What is remarkable about the foregoing scenarios are the purely instrumental value of spacepower to support terrestrial objectives, and yet their tremendous contribution drives the priority of early counter-space actions. Moreover, the nearness of the Earth and its power projection advantages means that it is like naval warfare taking place within constant range of shore-based artillery, or like artillery spotting balloons within range of the artillery. Describing such action as 'littoral' space warfare within the cosmic coastline is a helpful analogy. Sun Tzu also developed a taxonomy of different types of terrain and battle situations.[41] Sun Tzu might categorize

---

article/3847/1.

39. C. Priyant and Mark Surekha Kamath, "Review of Active Space Debris Removal Methods," *Space Policy* Volume 47, February 2019, Pages 194-206.

40. This is a reference to the prescriptions of Dr. Everett Dolman in *Astropolitik*.

41. Sun Tzu's six terrain configurations are: 1. Accessible (easily traversed both sides); 2. Entangling / suspended (hard to re-occupy; return impossible—attack

warfare in terrestrial space (approximately GEO and below as 'contentious ground' because of its great advantage to either side. However, he might also describe it as 'desperate' or 'hemmed-in ground' because these spaces provide no place of refuge, and ambush from Earth is easy.

In contrast, beyond GEO, space becomes 'accessible,' easily traversable by both sides, and relatively more easy to hide. But some places, such as the Lagrange points might be like narrow passes or 'constricted' ground, where if you are not the first to occupy, you should not enter. Other areas such as the poles of the Moon could be 'entangling ground,' proving hard or impossible to return or re-occupy once the enemy has made preparations, and the best one can do is lay distant siege, blockade or attack with fire.

An Expansionist View of Future Space Warfare: War Begins in Space

The expansionist view is that there is an extant scramble for the vast wealth of the inner solar system to win the geo-economic struggle[42]

---

if unprepared, don't attack if prepared; 3. Temporizing / stalemated (nobody gains by occupying); 4. Narrow passes/constricted (occupy first or don't enter); 5. precipitous heights; and 6. positions a great distance from enemy / Expansive. Sun Tzu's Taxonomy of Nine Battlegrounds / Situations / Terrain are: 1. dispersive ground (home territory; want to go home—fight not do not engage the enemy); 2. Facile/Light ground (slight penetration want to run back)—don't stop; 3. Contentious Ground (decisive point great advantage to either side—don't attack, hurry up my rear); 4. Open Ground/traversable (liberty of movement to both sides—don't block enemy, vigilant eye on defenses; do not allow your forces to become isolated); 5. Intersecting Highways /focal (easy control of 3 states—joint with allies; unite and form alliances with nearby lords); 6. Serious Ground/ Heavy terrain (deep penetration supply lines in danger—plunder and ensure continuous supplies); 7. Difficult Ground / entrapping terrain (forests, steeps, marshes hard to traverse—keep marching; move through quickly); 8. Hemmed in Ground / encircled terrain (easy to ambush with small numbers—resort to Strategem [trickery] use strategy, block any retreat); 9. Desperate Ground / fatal terrain (no place of refuge at all, saved from destruction only by fighting—Fight/ engage in battle). Adapted from: http://classics.mit.edu/Tzu/artwar.html.

42. Michael Brown, Eric Chewning, And Pavneet Singh, *Preparing The United*

for global hegemony. Space is not merely instrumentally important to national power, but is in fact the central theater of great power conflict.[43] As such, it is not only that a 'war could extend into space' but rather that jockeying for relative geo-strategic advantage in space constitutes its own spark of conflict, and escalation route.

Evolution of the Space Architecture and the Expected Order of Battle

The drive for space resources results in a different relevant force structure and different geostrategic pressures. The investments proceed along several lines of effort. The first is 'volume space access' and logistics, resulting in a tendency of all parties toward reusable and scheduled services. The second is in-space industrial capabilities such as robotic assembly,[44] 3-D printing,[45] vacuum spray

---

States for the Superpower Marathon With China, Brookings, April 2020, accessed June 30, 2020, https://www.brookings.edu/wp-content/uploads/2020/04/FP_20200427_superpower_marathon_brown_chewning_singh.pdf.

43. See in particular, Joshua Carlson, *Spacepower Ascendant: Space Development Theory and a New Space Strategy*, Independently Published via Amazon, June 27, 2020, https://www.amazon.com/Spacepower-Ascendant-Development-Theory-Strategy/dp/B08BWGPR8V/ref=tmm_pap_swatch_0?_encoding=UTF8&qid=1593376321&sr=8-1.

44. Mike Wall, "Incredible Technology: Spiderlike Robots Could Build Giant Space Structures," *Space.com*, April 06, 2015, accessed July 1, 2020, https://www.space.com/28846-spiderfab-space-structures-incredible-technology.html; Mike Wall, "Archinaut, a Construction Robot for Space, Could Launch a Test Flight in 2022," *Space.com*, July 13, 2019, accessed July 1, 2020, https://www.space.com/made-in-space-archinaut-flight-test-2022.html.

45. Lucie Gaget "3D printing in space: The next revolution?," *Sculpteo*, November 29, 2017 accessed July 1, 2020, https://www.sculpteo.com/blog/2017/11/29/3d-printing-in-space-the-new-revolution/; "Return to the moon? 3D printing with moondust could be the key to future lunar living," *The Conversation*, August 2, 2019, accessed July 1, 2020, https://theconversation.com/return-to-the-moon-3d-printing-with-moondust-could-be-the-key-to-future-lunar-living-121265; Bahar Gholipour, "3D Printing on Mars Could Be Key for Martian Colony," *Space.com*, October 03, 2013, accessed July 1, 2020, https://www.space.com/23059-3d-printing-mars-colony.html.

deposition,[46] etc. The third is resource prospecting and extraction, or 'space mining.' The final is a push for more capable in-space electrical power systems.

Desires to achieve scale for broader industrial ambitions seek markets with large and regular launch tempos such as satellite internet mega-constellations, suborbital[47] and orbital tourism,[48] commercial point-to-point transportation,[49] fuel depots,[50] and solar power satellites.[51] Pressures from industry move military constellations to be customers of in-space refueling,[52] in-space servicing,[53] and cargo

---

46. "Air Force Selects Lunar Resources and Rhea Space Activity for Next Generation USAF Space Capabilities," *PR Newswire*, October 31, 2019, accessed July 1, 2020, https://www.prnewswire.com/news-releases/air-force-selects-lunar-resources-and-rhea-space-activity-for-next-generation-usaf-space-capabilities-300949528.html.
47. Miriam Kramer, "Space tourism gets ready for launch," *Axios*, February 25, 2020 accessed July 1, 2020, https://www.axios.com/space-tourism-reshaping-industry-2bc34677-6fc1-4b5c-b81b-d1e0f69efcc4.html.
48. "Orbital Space Tourism," *Space Ventures Investors*, n.d., accessed July 1, 2020, http://www.spaceventuresinvestors.com/space-tourism.html#:~:text=was%20about%20300.-,Orbital%20Space%20Tourism,prototype%20orbital%20accommodation%20in%20orbit.
49. Dave Mosher, "Elon Musk: 'SpaceX is building floating, superheavy-class spaceports' for its Starship rocket to reach the moon, Mars, and fly passengers around Earth," *Business Insider*, June 16, 2020, accessed July 1, 2020, https://www.businessinsider.com/elon-musk-spacex-starship-ocean-spaceports-offshore-engineer-job-posting-2020-6.
50. John K. Strickland, Jr., "Orbital Propellant Depots: Building the Interplanetary 'Interstate Highway'," NSS, August 5, 2011, accessed July 1, 2020, https://space.nss.org/orbital-propellant-depots-building-the-interplanetary-interstate-highway/ ; "Why Build Big Rockets at All? It's Time for Orbital Refueling," *Universe Today*, August 16, 2019, accessed July 1, 2020, https://www.universetoday.com/143178/why-build-big-rockets-at-all-its-time-for-orbital-refueling/.
51. Department of Energy, "Space-Based Solar Power," March 6, 2014, accessed July 1, 2020, https://www.energy.gov/articles/space-based-solar-power.
52. NASA, "Robotic Tool Operations Bring In-Space Refueling Closer to Reality," August 16, 2019, accessed July 1, 2020, https://www.nasa.gov/feature/robotic-tool-operations-bring-in-space-refueling-closer-to-reality.
53. Mike Wall, "First-of-Its-Kind Satellite Servicing Spacecraft Launches on Russian Rocket," *Space.com*, October 09, 2019, accessed July 1, 2020, https://www.space.com/mev-1-satellite-servicing-spacecraft-launch-success.

delivery[54] to the Moon.

An awareness develops among the great powers / spacefaring states that the energy and material wealth of the solar system could decide geo-economic leadership[55] on planet Earth, and that space resources may be critical to their liberty and freedom of action. Occupation of key positions and control of global rule-sets offer leverage to control shares of this broader wealth and developing order.

The Centrality of the Lunar Poles: The 'peaks of eternal light'[56] surrounding the crater rims of the South and North Poles of the Moon assume particular relevance. While there is an abundance of Near-Earth Asteroids, the Moon is close, very large, and contains very large quantities of water[57] and methane ice which acts as a logistics force multiplier. Structural materials available by processing of the Lunar regolith (surface dirt) also offer a substantial logistics force multiplier as they can be placed into orbit at roughly 22x lower energy cost,[58] enabling construction of in-space industrial facilities or habitats of previously unheard-of scales. The possibility of bootstrapping (robots building robots), could enable exponential indus-

---

html#:~:text=MEV%2D1%20is%20the%20first,servicing%20craft%20to%20take%20flight.&text=The%20era%20of%20commercial%20satellite,Russian%20Proton%20rocket%20today%20(Oct.

54. NASA, "Commercial Lunar Payload Services," n.d., accessed July 1, 2020, https://www.nasa.gov/content/commercial-lunar-payload-services.
55. Peter Garretson, "The Future of the Asian Order Will Be Decided in Space: Space resources could determine which country winds up dominating the 21st century," *The Diplomat*, July 3, 2019, accessed June 29, 2020, https://thediplomat.com/2019/07/the-future-of-the-asian-order-will-be-decided-in-space/.
56. Martin Elvis, Tony Milligan, Alanna Krolikowski. "The Peaks of Eternal Light: A Near-term Property Issue on the Moon," (Postdoctoral NSF Grant#1066293), 2016. https://arxiv.org/ftp/arxiv/papers/1608/1608.01989.pdf.
57. Loren Grush, "Why mining the water on the Moon could open up space exploration: 'Water is the oil of space'," *The Verge*, August 23, 2018, accessed July 1, 2020, https://www.theverge.com/2018/8/23/17769034/nasa-moon-lunar-water-ice-mining-propellant-depots.
58. Mark Evan Prado, "Moving around in the Earth-Moon system," *Permanent.com*, accessed June 29, 2020, https://www.permanent.com/space-transportation-earth-moon.html#Earth-Moon-energy.

trial growth. The possibility of building a global system of orbital power plants (and potentially controlling close to 10% of the global economy) offers a market of astounding proportions.

Space Lines of Commerce: What emerges are space lines of commerce of geo-economic significance, or at least perceived as the critical pathways which imply future geo-economic significance and must be protected. The Moon becomes the source of raw materials and location of some industrial processing. People travel to and return from the Moon from LEO to Lunar orbits. Supplies which support industrial processes ('vitamins') flow from Earth to the Moon, while high-mass structural materials ('macro-nutrients') or finished goods flow from the Moon to various staging or profit centers, including lunar orbits, L1, L2, GEO, LEO, L4 and L5.[59] In this period, the primary profit centers remain in GEO and LEO.

Military Posture: Space militaries are called to assume a broader set of responsibilities, more similar to navies and coast guards. As civil and commercial operations begin in Cis-Lunar space beyond GEO, there is increased pressure to provide safety-of-navigation services: navigational aids, space domain awareness, rescue and tug services, active debris removal, safety inspections, and patrol.[60] Pressure also exists to base military patrol and logistics craft based upon commercial counterparts. While generally cooperative, space militaries and their owning states constantly measure their posture vs one another. States may be averse to openly weaponizing, but capabilities required to perform safety-of-navigation services are dualuse and inherently have space control capabilities.

---

59. For an in-depth discussion of the Cis-Lunar 'Celestial Lines of Commerce' see Sean McClain, *Celestial Sentinels*.
60. See for example: Brent Ziarnick, "the US Space Guard: Institutional Support to Space Commerce" Chapter 20 in Langdon Morris and Kenneth Cox (editors) *Space Commerce: The Inside Story by the People who are Making it Happen* (Aerospace Technology Working Group, 2010) pg 351; M.V. Smith, The Space Corpsman's Handbook, unpublished, January 1, 2019; Sean McClain, *Celestial Sentinels: A Framework For Cis-Lunar Space Domain Awareness In 2035*, Air Command and Staff College (Air University), March 2020.

Expansionist Scenarios for Space Warfare

The logic of these situations is largely consistent whether we are talking about the very first stages of mineral exploration with robotic landers and rovers or large-scale industrial activity and settlements. What is different about these scenarios is that they *begin* over equities in space, and military action takes place primarily in space. Because the interests at stake are in space, there is at least the possibility of space forces acting unilaterally and that such action by space forces could prove 'decisive' for the encounter.

Interdicting 'Unacceptable' Behavior: With people in space come questions of political expression, symbolism and different standards of behavior. There are any number of scenarios where the mere presence of a symbolically important person or group, or their activities could incentivize a rival power to interdict their communications or movement. Things that might not seem problematic to the US, might appear problematic to another power such as the PRC. The presence of religious rights activists (such as the Dalai Lama[61] or Fao Lung Gong[62]), or political dissidents could use in-space locations to militate for change, such as using a private space station for broadcast.[63] This could create interdiction of communications (jamming) or logistics (blockade) which could be seen as unacceptable. Other situations could involve the emplacement of capabilities deemed by the other to represent weapons or fortifications in disallowed zones, (for example: placing a mining laser on the Moon which can point

61. Matthias Williams, "India rejects Chinese protest at Dalai Lama trip," *Reuters*, September 30, 2009, accessed June 29, 2020, https://in.reuters.com/article/india-china-dalai-lama/india-rejects-chinese-protest-at-dalai-lama-trip-idINDEL53581420090930.
62. Rick Rojas, "Falun Gong followers protest Chinese President Xi in desert visit," *Los Angeles Times*, June 8, 2013, accessed June 29, 2020, https://www.latimes.com/local/lanow/la-me-ln-falun-gong-xi-protest-20130607-story.html
63. For example, see the case of the PRC's strong-arm tactics to prevent the Voice of America from airing an interview of political dissident, Chinese billionaire Guo Wengui. Evelyn Cheng, "How an interview with one Chinese billionaire threw a US broadcaster into turmoil," *CNBC*, June 9, 2017, Accessed June 27, 2020, https://www.cnbc.com/2017/06/09/interview-with-guo-wengui-throws-voice-of-america-into-turmoil.html.

up, or a mass driver that could direct-fire toward Earth). If one side became aware of an attempt to emplace an operational capability for either anti-satellite capabilities or Earth-strike capability on the Lunar surface, the other side would almost certainly pre-empt via embargo (similar to the Cuban missile crisis[64]), sabotage (like Stux-Net[65]), or attack (such as the Israeli Operation Opera air attack on Iraqi nuclear facilities[66]).

Competing Claims: Because national interests are geo-economic, corporate interests are likely to be perceived as national interests. If disagreements arise about operating areas, safety, or harmful practices which interfere with another state's corporate interests, disagreements could arise. A non-state actor such as a corporation could impound or exclude other parties, or could convince their sponsoring state to do so. They might arrest or hold hostage people, equipment, supplies or valuables.

Sovereign Claims: A special case of both of the above would involve behaviors interpreted by the other as constituting a sovereign claim or appropriation, such as the announcement of a broad Exclusive Economic Zone,[67] a Defense Identification Zone,[68] or anything considered to be an excessive land grab or land grab of a unique

64. Editors, "Cuban Missile Crisis," *History.com*, January 4, 2010, accessed June 29, 2020, https://www.history.com/topics/cold-war/cuban-missile-crisis.
65. Kim Zetter, "An Unprecedented Look at Stuxnet, the World's First Digital Weapon," *Wired*, November 3, 2014, accessed June 29, 2020, https://www.wired.com/2014/11/countdown-to-zero-day-stuxnet/.
66. Anthony D'Amato, "Israel's Air Strike upon the Iraqi Nuclear Reactor," Faculty Working Papers (2010), Paper 76. http://scholarlycommons.law.northwestern.edu/facultyworkingpapers/7.
67. Christopher Woody, "This new Defense Department map shows how China says one thing and does another with its military operations at sea," *Business Insider*, August 17, 2018, accessed June 29, 2020, https://www.businessinsider.com/defense-department-map-shows-chinas-naval-operations-in-maritime-eezs-2018-8.
68. Jun Osawa, "China's ADIZ over the East China Sea: A 'Great Wall in the Sky,'" Brookings, December 17, 2013, accessed June 29, 2020, https://www.brookings.edu/opinions/chinas-adiz-over-the-east-china-sea-a-great-wall-in-the-sky/.

and special area. Although restrained claims of use and associated safety zones are likely to be tolerated, certain actions which might significantly alter the status quo and create a significant precedent or advantage might be interpreted as a failure of the treaty and necessitating self-defense.

Atrocity: Any of the above could involve or provoke some kind of atrocity by local actors. For example, corporate actors angry over some trespass might kill their competitors directly or do so via sabotage. Angry nationalists could do the same in response to the other side hosting dissidents.

Coercion & Counter-Coercion: In all such cases, there will be an attempt to compel behavior and likely counter-action to posture to compel its relief or to liberate-by-force. A crisis will emerge where one state attempts to compel the other by creating a local superiority and signaling its seriousness, and the other side challenges that attempt at coercion. This may occur because of the geo-economic or positional stakes which would cede too much advantage to the other, or this may occur 'on principle' where some action is symbolic of an attack on sacred values, commitments or identity.

Therefore, we are likely to see military assets move into positions where they can threaten 'misbehaving' actors with direct force or interdiction of supplies, or to wage astroeconomic warfare by interdicting outbound commerce or threatening capital assets or profit centers to impose costs and force the offending side to capitulate. The objecting nation will move to counter-threaten the coercing fleet and/or counter-threaten a like asset as mutual hostage-taking.[69]

69. "capture" is a neglected consideration in air and spacepower thinking. Various forms of "active control" including Defeat, Capture, Containment, and Conversion are discussed in Jeremy L. Renken, "Airpower for Irregular Warfare: Reconciling USAF Theory and TACAIR Design with the Demands of Irregular Warfare and Special Operations," (master's thesis), National Defense Intelligence College, July 2012.

Scenarios: Warfare

While there is a constant background of competitive posture between militaries which position, inspect, and maneuver akin to Cold War submarine operations[70] -- which could result in 'dangerous activities at sea' -- this is not the primary route to escalation.

Incentives to 'Go Big': Both sides are likely to 'wish the other weren't there' seeing the adversary as a nuisance in daily operations, and feeling constantly on-guard that they will overstep 'appropriate' boundaries, or are just waiting for an opportunity to eliminate the other. Therefore, it seems plausible that both sides will incentive to develop war plans to decisively eliminate the other's in-space military capability and to wage astroeconomic warfare by holding the adversary's commerce and in-space profit centers at risk.

Counter-Force and Counter-Value: In these scenarios, military counter-force is concerned with removing all elements of military presence and force projection which can be used to coerce, raid or destroy commerce. Counter-value includes all portions of the in-space industrial and space-services infrastructure and value chain. A belligerent might target Lunar mines, mining equipment, launch infrastructure, propellant infrastructure, factory infrastructure, profit generating infrastructure within Terran space (GEO and below)

70. Jon Bowermaster, "The Last Front of the Cold War: You think the Cold War is over? Think again. Russian and American forces are still challenging each other in the Arctic," *The Atlantic Online*, November 1993, accessed June 29, 2020, https://www.theatlantic.com/past/docs/politics/foreign/front.htm; Owen R. Cote, Jr., "Cold War Struggle with Soviet Submarines," FAS, March, 2000, accessed June 29, 2020, https://fas.org/man/dod-101/sys/ship/docs/cold-war-asw.htm.

such as satellite power plants,[71] data processing centers,[72] communication platforms[73] and human habitats.[74] Although conflict could begin over geo-economic or geo-symbolic considerations in GEO or LEO and might follow similar escalation routes, the most interesting scenarios take place in deep space.

Deep Space Considerations: The likely most important terminus of deep space commerce is the Moon. On the Moon itself are major resources, key positions, and potential fortifications. Unlike space itself, it may be possible to hide or harden on the Moon. As the geo-economic and geo-strategic hub of deep space transportation and commerce,[75] the Moon seems the most likely case to explore deep-

71. Kirk Nankivell , "Why the Future of Solar Power Is from Space," *Singularity Hub*, Dec 31, 2018, accessed July 1, 2020, https://singularityhub.com/2018/12/31/why-the-future-of-solar-power-is-from-space/ ; Daniel Oberhaus, "How to Get Solar Power on a Rainy Day? Beam It From Space," *Wired*, December 2, 2019, https://www.wired.com/story/how-to-get-solar-power-on-a-rainy-day-beam-it-from-space/ ; Eric Rosenbaum and Donovan Russo, "China plans a solar power play in space that NASA abandoned decades ago," *CNBC*, March 17, 2019, accessed July 1, 2020, https://www.cnbc.com/2019/03/15/china-plans-a-solar-power-play-in-space-that-nasa-abandoned-long-ago.html.
72. DSM, "Is the Future of Data Centers in Space?," *DSM*, November 6, 2018, accessed July 1, 2020, https://www.dsm.net/it-solutions-blog/is-the-future-of-data-centers-in-space ; Michael Sheetz, "Satellite start-up raises $100 million to put cloud data storage in space," *CNBC*, December 20 2018, accessed July 1, 2020, https://www.cnbc.com/2018/12/19/cloud-constellation-raises-100-million-to-store-cloud-data-in-space.html.
73. B. Edelson, R. Lovell and C. Cuccia, "The Evolution of the Geostationary Platfom Concept," *IEEE Journal on Selected Areas in Communications* (Volume: 5 , Issue: 4 , May 1987).
74. Hanneke Weitering, "6 Private Deep Space Habitat Concepts That May Pave the Way to Mars," *Space.com,* October 11, 2016, accessed July 1, 2020, https://www.space.com/34357-private-deep-space-habitat-concepts-for-nasa.html ; Kate Baggaley, "Huge Space Hotel Promises Fake Gravity and 'Supersized Basketball: The Planned Von Braun Station Could be Open as Soon as 2027," *NBC News*, accessed July 1, 2020, https://www.nbcnews.com/mach/science/huge-space-hotel-promises-fake-gravity-supersized-basketball-ncna1051596.
75. Dennis Wingo, *MoonRush: Improving Life on Earth with the Moon's Resources* (Ontario, Canada: Apogee Books, 2004); Paul D. Spudis, *The Value of the Moon: How to Explore Life and Prosper in Space Using the Moon's Re-*

space warfare in this period.

Significant choke points of Cis-Lunar commerce occur at EML1, Low Lunar Orbit, and the Lunar Poles. Traffic to and from the Moon likely will flow through these points. Therefore, to interdict positions, it is desirable to hold these positions. The difficulty of holding these positions depends on a variety of considerations. The viability of holding EML1 and LLO depend a great deal on whether there are ASAT capabilities on the Moon itself, and of what strength. If the Moon has a developed industrial capability, it might be able to hold at risk anything in its near abroad. On the other hand, if the Moon is undefended and its industrial base concentrated, then embryonic defenses may be crushed with conventional weapons from orbit, and such attacks would not be illegal under the OST since the arms are not stationed on the Moon and are not weapons of mass destruction.

Unlike in the earlier scenario, military space activities may no longer be dominated by the overwhelming influence of the Earth. Local advantage and long-distance power projection is likely to matter, *and fait accompli's* are possible. The Moon is approximately ten-times more distant than GEO. This, along with light pollution from the Moon's brightness mean that effective surveillance likely require local sensors. Efficient transit between the Earth and the Moon requires approximately three days and significant energy expenditure. Therefore, repositioning or attack may not be timely. Defenders in the vicinity of the Moon may have warning and time to prepare. At some point, an industrial and logistical capability on the Moon may allow superior local logistics. The vastness of Cis-Lunar space combined with weak 3-body gravitational interactions allow significantly increased possibilities for maneuver, hiding and surprise by military forces.

Space Cruisers: Very likely the vessels that service this area of space are general purpose and multi-use. They might have dedicated armaments such as guided missiles, rail-guns, or directed energy

---

sources (Washington, DC: Smithsonian Books, 2016).

weapons, but they need not. Perhaps they have only robotic effector arms or merely the ability to maneuver and therefore block. Almost certainly they will have some degree of SSA and autonomy, whether robotic or manned. Therefore, space warfare in this theater might most resemble naval battles at the far-reaches of empire.

Privateers, Reserve Forces & Preventing Self-Help: Of note, assets brought to bear might also be 'merchant marine' or repurposed commercial construction and transportation forces, perhaps even including authorization to act as privateers or private security firms (privateers). In fact, if states choose to focus on conservative scenarios as described in the previous section and neglect to maintain a presence apace with the private sector, the entire scenarios described below could in principle take place entirely between non-state actors using assets principally developed for industrial purposes. A key reason why states might wish to develop Cis-Lunar forces is to prevent a tendency of corporations toward 'self-help.'

Sequence of Events: The initiating coercer will move forces into the area to deny transit and directly threaten the locations in contest. Whatever was the initial balance, we can imagine that events might unfold as follows:

1. The coercer will principally move forces to apply direct coercion against the target and posture other forces to check a would-be responder.

    a. This may be in space, but it may also include actions on the Lunar surface, such as driving a robotic rover to block an opponent's landing pad or airlock.

2. The responder is likely to counter position forces against the coercer's forces (a vertical escalation) and the coercer's other forces (horizontal escalation).

3. If space remains an offense-dominant environment, there will be a strong incentive to pre-empt to enjoy the first mover advantage.

4. Each side will likely threaten each other's dedicated SDA assets

to reduce the local SDA.

5. Each side will 'point their guns' at the most threatening asset—the assets which either most directly threaten their interests or which have the greatest capacity to create military losses.

6. GEO and LEO become additional theaters in their own right.

7. As long as both sides feel they can achieve greater advantage in waiting, both sides will reinforce and horizontally escalate.

   a. This may certainly include horizontal escalation on planet Earth itself, and not only against space infrastructure, but anything of value which may be threatened in trade.

   b. Both sides will be tempted to pre-empt or blockade further forces from leaving Earth to alter the balance of forces.

8. At some point in time the drawing up of forces will culminate for one power or the other, with it becoming clear that any delay merely puts them in a worse position.

9. At that point (in a disciplined situation), one side initiates conflict. At any point along this continuum of posturing, an individual asset might decide it must act in self-defense and trigger the same cascade.

   a. Unless the situation is extremely desperate, very likely the coercion itself will be neglected until the decision of battle.

   b. Because each side must respect the destructive power of the other, those platforms become the focus of effort to ensure follow-on freedom of action.

   c. If they are close enough to sense and shoot each other autonomously, then destruction of enemy reconnaissance and interdiction of logistics or follow-on forces become supporting efforts.

   d. If a principle threat is on the Lunar surface, it will either be struck immediately or in-space forces will remain distant enough to give them increased chances of survival.

e. Whether or not capital assets such as a propellant depot or lunar fueling and landing facilities are struck depend a great deal on several factors: do they materially contribute to the extant battle? Will their destruction cause harm to one's own interests? For example, they might foul orbits or puncture one's own facilities. Can they be captured either to be held hostage to exact future concessions or to add to one's forces?

10. Either through attrition or paralysis one side will gain sufficient advantage to remove the other or put them in such an un-tenuous position as to negotiate a cease fire.

Direct-Fire vs Swarms: Long-range direct-fire weapons within the theater are likely to be the primary initial focus of action because of their reach and deep magazines. Very likely their destruction will be affected by a combined arms approach involving swarms of attackers overwhelming its self-defense capacities. The principle early threat to all platforms will be co-orbital 'snugglers'[76] which would need to be 'de-loused' early. The realities of orbital maneuver are likely to mean that the timing of such operations is unlikely to be well coordinated in time, leading to significant fog and friction.

Restraint & Unrestraint

Restraint: States will likely wish to avoid the audience costs of being seen either as the aggressor or as weak. Although locations on and around the Moon are likely to have significant strategic value, belligerents will wish to avoid overt military conflict unless they can clearly assert 'self-defense.' They may also prefer to steer clear of conflict to exploit absolute gains, especially if they believe that the time to pursue relative gains is yet in the future.

76. Brian Chow, "China's new space threat and the justification of US pre-emptive self-defense," *The Space Review*, January 18, 2016, accessed July 1, 2020, https://www.thespacereview.com/article/2903/1.

The problem of also maintaining supply lines for one's personnel at a lunar facility would also loom large, and perhaps constrain the barbarism or 'salt the Earth' strategies that are possible.

Unrestraint: The possibility of a wide-scale attrition of one's competitor leaving one in a consolidated and strong position would create an incentive toward broad scale offensive.

The considerations of restraint when one has isolated personnel would go away if they were eliminated, and could result in an escalation to reprisal or fouling of the local environment through area-weapons and/or large amounts of space debris.

Dialectic Response

Several outcomes are possible which would condition the system for future geo-political conflict:

Coercer Achieves Small Wins: If the coercer's actions in crisis or limited war succeed in appeasement, this conditions further appeasement, likely to result in further 'salami-slicing' through successive coercions which change the facts-on-the-ground in favor of the coercer. Such actions however will create resentments, setting the stage with tectonic forces for the next confrontation.[77]

Coercer Achieves Big Wins: If the coercer wins in a wider war, it is now likely to exclude its adversary from its closed economic empire, or to extract such concessions as to only allow it the weakest of interactions.[78]

Restrained Responder Wins: If the winning power had up until the battle been restrained in its weaponization and expression of military spacepower, it is likely to see such unpreparedness as unaccept-

77. Examples include the subordination of Germany after WW1, or the counter-reaction to appeasement after Nazi Germany successively took territory.
78. An example of such an outcome is illustrated in Worden & Shaw's Asian Hegemony scenario beginning page 57.

able. Convinced that the adversary has a demonstrated character as an aggressor, and the taboo of space warfare having been broken, the winning power, previously restrained, is likely to go 'full Dolman' and attempt to establish and maintain military dominance and overwhelming presence in the deep space Area of Responsibility (AOR).

Implications

The significant divergence between the conservative and expansionist scenarios suggest that the United States might be preparing, planning and equipping for the wrong war, or at least not the full spectrum of possible wars. This could leave the United States, its allies and their respective businesses open for coercion. Preparation for the full spectrum of space crisis and conflict requires early attention to military doctrine, concept design, force structure design, and international governance including arms control and the law of armed conflict, as well as for legislation and policy.

Military Doctrine: Military doctrine must anticipate that United States geo-strategic interests will likely encompass economic activities in deep space, and develop concepts for peacetime strategic offensives to maintain positions of advantage.

Concept Design: Conceptual design of military campaigns and crisis planning need to anticipate the full range of scenarios to include the distant defense of geo-strategic economic, symbolic, positional and citizen interests, and consider how such counter-coercion and defense would take place.

Force Structure Design: Military space vehicles capable of defending US interests in the vicinity of the Moon differ in important ways from satellites designed to support terrestrial forces. Design of a 2030-2040 force structure requires attention to the unique navigational, maneuver, logistics, and power projection needs for such vehicles to be effective.

International Governance including Arms Control and the Law of Armed Conflict: Consideration and anticipation of conflict may enable the community of nations to arrive at consensus on certain conflicts or types of conflict they mutually wish to avoid. As a clearer picture can be painted of harm to be avoided, there may be opportunities to: avoid certain conflicts through bilateral or multi-lateral agreements which avoid ambiguity of claims and provide forums to adjudicate would be conflicts; make certain acts or weapons illegal; agree to limit the type and number of certain arms which could impede broader interests. In particular, prospects for restraint may be more likely if there is an early conditioning shock that results in sufficient horror and few benefits. Such shocks might include a limited ASAT exchange between states or a rouge state or actor attempting to hold the world hostage or exact costs through a disproportionate or terror weapons such as a Electro-magnetic Pulse (EMP), High Altitude Nuclear Detonation (HAND), or deliberate collision.

Legislation and Policy: Because the conservative view is so firmly established in the US Department of Defense, there is a significant possibility that left to its own devices, it will be underprepared for expansionist scenarios. This suggests a role for civilian leadership to ensure the US DoD is prepared through legislation and policy to specify roles and missions to protect space commerce, to specify plans, and to specify the Area of Responsibility (AOR) as encompassing the entirety of the Cis-Lunar theater.

Conclusion

In this essay, we have explored two visions for space warfare which might manifest in the 2030-2040 timeframe. The conservative view postulated scenarios where war begins over Earth-bound geostrategic interests but extends into space to maintain advantage for the terrestrial warfighter. The expansionist scenarios anticipate that nations may have reason to fight over geo-strategic interests resident in space, and over equities relating to Lunar commerce in particular. Circa 2030 (and perhaps even sooner), as the economic development activities in deep space establish geo-strategic interests on other planetary bodies (starting with the Moon), the possibility for the expansionist view of crisis and conflict becomes more likely. While some may wish to dismiss the expansionist view, they should consider that even at the time of writing, 10-20 years in advance of the scenarios, a number of nations are already sending landers to the Moon (United States, China, India, Russia, Japan) as precursors to economic development. Being prepared for the full spectrum of space crisis and conflict requires early attention to military doctrine, concept design, force structure design, and international governance including arms control and the law of armed conflict, as well as for legislation and policy.

APPENDIX

This appendix explores in greater detail the factors which condition the specific shape of space conflict.

Why Consider Space Warfare?

Since the 2007 Chinese ASAT test,[79] there has been significant public interest in threats to space systems. With the formation of China's Strategic Support Force,[80] the subsequent formation of a USSPACECOM[81] and US Space Force[82] and corresponding assertion by the US that its adversaries had transformed space into a warfighting domain, there is broad interest regarding what a 'space war' might look like.

The essay is of course a speculative work. In it, I have attempted to explore the trends and causal pressures which might shape conflict in space. I have no crystal ball. None of us can know the future—tremendous surprises have occurred in a span of two decades. Moreover, *at times the role of speculation is sometimes not to predict but to prevent*, because foresight offers early feedback loops to shape present choices. The value of foresight work is to anticipate the range of possible futures to reduce surprise to enable early rec-

---

79. Leonard David, "China's Anti-Satellite Test: Worrisome Debris Cloud Circles Earth," *Space.com*, February 02, 2007, accessed June 29, 2020, https://www.space.com/3415-china-anti-satellite-test-worrisome-debris-cloud-circles-earth.html.
80. John Costello and Joe McReynolds, *China's Strategic Support Force: A Force for a New Era, Center for the Study of Chinese Military Affairs*, Institute for National Strategic Studies China Strategic Perspectives, No. 13, NDU Press, October 2019, accessed June 30, 2020, https://ndupress.ndu.edu/Portals/68/Documents/stratperspective/china/china-perspectives_13.pdf.
81. DOD, "Department of Defense Establishes U.S. Space Command," *Defense.gov*, August 29, 2019, accessed June 30, 2020, https://www.defense.gov/Newsroom/Releases/Release/Article/1948288/department-of-defense-establishes-us-space-command/.
82. USSF, "About U.S. Space Force," n.d., accessed June 30, 2020, https://www.spaceforce.mil/About-Us/About-Space-Force.

ognition and adaptation.

Literature Review

Although there is a diversity of spacepower theory and taxonomies by which to divide them, for the purposes of this paper, the conservative view of spacepower and the expansionist view of spacepower are roughly equivalent to the main divergence in current spacepower thought, between the 'brown water thinkers' and the 'blue water' thinkers, [83] each of which emphasize different theaters of competition and conflict.

'Brown Water' Schools of Spacepower: On the one side are the 'brown water' thinkers, they see the primary analogy as continental seapower. Particularly coherent are Brad Townsend[84] and Bleddyn E. Bowen[85] who have explored space warfare within Earth orbital space where Earth-based counter-space is a dominant influence. They focus principally on how military spacepower supports terrestrial conflict. Paul Szymanski[86] also appears to fall primarily in this camp, as does John Klein.[87]

---

83. The distinction between Blue and Blown Water schools of space power was first advanced by Dr. Brent Ziarnick: Brent D. Ziarnick, "The battle for the soul of the Space Force," *The Hill*, August 23, 2019, accessed July 1, 2020; Brent D. Ziarnick, "The Space Force strategic vision emerges," *The Hill*, September 9, 2019, accessed July 1, 2020 https://thehill.com/opinion/national-security/460469-the-space-force-strategic-vision-emerges.
84. Brad Townsend, "Space Power and the Foundations of an Independent Space Force," *Air & Space Power Journal*, Winter 2019, accessed June 10, 2020, https://www.airuniversity.af.edu/Portals/10/ASPJ/journals/Volume-33_Issue-4/F-Townsend.pdf.
85. Bleddyn E. Bowen, *War in Space: Strategy, Spacepower and Geopolitics*, (Edinburgh: Edinburgh University Press, 2020).
86. Paul Szymanski's various products can be found in his online collection here: http://www.space.forces.space/.
87. John Klein, *Space Warfare: Strategy, Principles, and Policy* (London & New York: Routledge, 2006); John Klein, *Understanding Space Strategy: The Art of War in Space (Space Power and Politics)* [London & New York: Routledge,

'Blue Water' Schools of Spacepower: On the other side are the 'blue water'[88] thinkers who have advanced theories of how the space domain itself can be used in the contest for political primacy, using the analogy of the deep blue oceans which carry trade. These thinkers include Carl Everett Dolman,[89] Simon "Pete" Worden & John Shaw,[90] Brent Ziarnick,[91] Chris Stone,[92] and Joshua Carlson.[93]

Avoiding Those Conflicts: While not a school of military spacepower per se, a third school of thinkers spend their energies attempting to avoid conflict. These are the arms controllers. They are less concerned with the conduct of space war but rather develop thinking on the inherent danger of space conflict to terrestrial equities, and are associated with arms control or environmental security. These include, Joan Johnson-Freese,[94] Clay Moltz,[95] and Dan Deudney.[96] A key difference between arms controllers and other schools is the confidence with which arms controllers project the total negative costs and the confidence they have that rational actors should pursue optimal decisions before either crisis or capability emerges. In

---

2019].

88. Dr. Brent Ziarnick is the first to call this collection of thinkers "the blue water school of spacepower"; Dr. Simon "Pete" Worden, Brig Gen, USAF-Ret was the first to my knowledge to assert US space forces must focus on blue water objectives.

89. Everett C Dolman., *Astropolitik: Classical Geopolitics in the Space Age.*

90. Simon P Worden and John E. Shaw, *Whither Space Power? Forging a Strategy for the New Century* (Maxwell Air Force Base: Air University Press, 2002).

91. Brent Ziarnick, *Developing National Power in Space: A Theoretical Model.* Jefferson (NC: McFarland & Company, 2015).

92. Christopher M. Stone, *Reversing the Tao: A Framework for Credible Space Deterrence* 1st Eds (Scotts Valley, California: CreAte Space Independent Publishing Platform, 2016).

93. Joshua Carlson, *Spacepower Ascendant*, (Kindle Self-Publish, 2020).

94. Joan Johnson-Freese, *Space as a Strategic Asset* (New York: Columbia University Press, 2007).

95. James Clay Moltz, *Asia's Space Race: National Motivations, Regional Rivalries and International Risk.*

96. Daniel H Deudney, *Dark Skies: Space Expansionism, Planetary Geopolitics, and the Ends of Humanity*. New York: Oxford University Press, 2020.

contrast, the other two schools of thought question the certainty of absolute losses (holding out positive outcomes for one side either in warfare or coercion), and assert the irrationality of restraint, saying "nobody does arms control with somebody who doesn't have the arms."[97]

Visions of Space Warfare

I am not alone in imagining space conflict. The reader can find other imaginative scenarios for conflict or space warfare treated in non-fiction, including Worden & Shaw's *Whither Space Power,*[98] Fred Kennedy's "A failure of vision,"[99] Charlie Dunlap's "How we lost the high tech war of 2007,"[100] George Freedman's *The Next 100 Years,*[101] Andrew Krepinevich's *7 Deadly Scenarios,*[102] Tangredi's *All Possible Wars*[103] *and OECD's Geopolitical Developments and the Future Of The Space Sector.*[104] While aimed principally at applying a geopolitical determinist theory to space, Dan Deudney's Dark Skies[105] provides a thoughtful examination of a number of space

97. My thanks to Lt Col Jeremy Renken who made me aware of this phrase.
98. Simon P Worden and John E. Shaw, *Whither Space Power? Forging a Strategy for the New Century* (Maxwell Air Force Base: Air University Press, 2002).
99. Fred Kennedy, Rory Welch, Bryon Fessler, "A Failure of Vision Retrospective," *Airpower Journal*, Summer 1998, accessed June 10, 2020, https://apps.dtic.mil/dtic/tr/fulltext/u2/a530286.pdf.
100. Charles J. Dunlap, Jr, "How We Lost The High-Tech War Of 2007: A Warning from the Future," *The Weekly Standard*, January 29, 1996, accessed June 10, 2020, https://www.washingtonexaminer.com/weekly-standard/how-we-lost-the-high-tech-war-of-2007.
101. George Friedman, *The Next 100 Years: A Forecast for the 21st Century* (Anchor, 2010).
102. Andrew Krepinevich, *7 Deadly Scenarios: A Military Futurist Explores the Changing Face of War in the 21st Century* (NY: Bantam, 2010).
103. Sam J. Tangredi, *All Possible Wars? Toward a Consensus View of the Future Security Environment, 2001–2025*, Institute for National Strategic Studies, National Defense University, Washington, D.C., 2000.
104. Michel Andrieu, *Geopolitical Developments and the Future Of The Space Sector,* OECD, Friday, 7th May 2004.
105. Daniel H Deudney, *Dark Skies: Space Expansionism, Planetary Geopoli-*

conflict scenarios.

Fiction written specifically to offer strategic insight has also examined space warfare, including Peter Singer's *Ghost Fleet,*[106] and William Scott's *Space Wars*,[107] and *Counter Space.*[108] The most comprehensive and imaginative treatment of space warfare would be found in the corpus of science fiction literature. Such a survey of space warfare as depicted in science fiction is beyond the scope of this work, but is something which would be an instructive undertaking by organizations concerned with space warfare.

What Is Known about Space Weapons Development?

Space Weapons of great variety were explored in the Cold War, especially during the Strategic Defense Initiative (SDI) time period,[109] after which, in the 'unipolar moment'[110] strategic thinking about space weapons atrophied, and the United States treated space as a sanctuary. As it became clear US conventional military power was reliant on space, and that space was a sort-of unprotected rear area for the US, states whose vision of the global order diverged from the

tics, and the Ends of Humanity. New York: Oxford University Press, 2020.

106. P. W. Singer and August Cole, *Ghost Fleet: A Novel of the Next World War* (New York: Eamon Dolan/Mariner Books, 2016)

107. Michael J. Coumatos, William B. Scott, William J. Birnes, *Space Wars: The First Six Hours of World War III, a Wargame Scenario* (Forge Books, 2010).

108. William B. Scott, Michael J. Coumatos, William J. Birnes, *Counterspace: The Next Hours of World War III* (Forge Books, 2009).

109. See for example, Collins, John M. *Military Space Forces: the next 50 years*. Vol. 4. Brassey's Incorporated, 1989, available online at: https://www.spacefaringamerica.com/wp-content/uploads/2020/05/military_space_forces_collins_1989_csr_report.pdf; Preston, Robert, Dana J. Johnson, Sean JA Edwards, Michael D. Miller, and Calvin Shipbaugh. *Space Weapons Earth Wars*. RAND Corporation, 2002.

110. Charles Krauthammer, "The Unipolar Moment: America and the World 1990," *Foreign Affairs*, September 18, 1990, accessed June 30, 2020, https://www.foreignaffairs.com/articles/1990-01-01/unipolar-moment#:~:text=The%20unipolar%20moment%20means%20that,now%20in%20the%20Persian%20Gulf.

US began to develop anti-satellite weapons, most notably China and Russia. A number of public threat assessments have described both the general state of affairs[111] as well as specific adversarial threats.[112] They detail broad lines of anti-satellite and counter-space efforts known to be explored by adversaries of the United states including threats electronic warfare (EW) such as jamming, ground-launched KE-ASATS (even able to reach GEO), directed energy weapons such as high-power lasers, cyber-attacks, and ground-launched high-altitude nuclear detonations (HAND), as well as orbital threats including radio frequency jammers, kinetic kill vehicles (KKVs), chemical sprayers, robotic mechanisms, high power microwaves (HPM), and Lasers.[113] Moreover, many lines of counter-space ef-

---

111. Todd Harrison, Kaitlyn Johnson, Thomas G. Roberts, Tyler Way, and Makena Young, *Space Threat Assessment 2020*, CSIS, March 2020, accessed June 30, 2020, https://csis-website-prod.s3.amazonaws.com/s3fs-public/publication/200330_SpaceThreatAssessment20_WEB_FINAL1.pdf ; Brian Weeden and Victoria Sampson, *Global Counterspace Capabilities: An Open Source Assessment*, Secure World Foundation, April 2020, accessed June 30, 2020, https://swfound.org/media/206970/swf_counterspace2020_electronic_final.pdf
112. NASIC, *Competing in Space*, December 2018, accessed June 30, 2020, https://media.defense.gov/2019/Jan/16/2002080386/-1/-1/1/190115-F-NV711-0002.PDF ; Defense Intelligence Agency, *Challenges To Security In Space*, January 2019, accessed June 30, 2020, https://www.dia.mil/Portals/27/Documents/News/Military%20Power%20Publications/Space_Threat_V14_020119_sm.pdf ; Mark Stokes, Gabriel Alvarado, Emily Weinstein, and Ian Easton, *China's Space and Counterspace Capabilities and Activities* (Prepared for: The U.S.-China Economic and Security Review Commission), March 30, 2020, accessed June 30, 2020, https://www.uscc.gov/sites/default/files/2020-05/China_Space_and_Counterspace_Activities.pdf; Eric Heginbotham, Michael Nixon, Forrest E. Morgan, Jacob L. Heim, Jeff Hagen, Sheng Tao Li, Jeffrey Engstrom, Martin C. Libicki, Paul DeLuca, David A. Shlapak, David R. Frelinger, Burgess Laird, Kyle Brady, and Lyle J. Morris, *The U.S.-China Military Scorecard Forces, Geography, and the Evolving Balance of Power*, 1996–2017, RAND, 2015, p. 227-241, https://www.rand.org/content/dam/rand/pubs/research_reports/RR300/RR392/RAND_RR392.pdf.
113. As broad as this list is, early Cold War thinking (before the Outer Space Treaty of 1967 and Partial Test Ban treaty) was far more expansive, including strategic orbital bases, orbital and suborbital space bombers capable of nuclear and strategic strikes, Moon bases, manned orbiting laboratories, fully-reusable spaceplanes for logistic access, and nuclear thermal and nuclear pulse-powered

fort may be concealed or justified under dual-use purposes, such as Active [Space] Debris Remediation (ADR), Satellite Servicing, or experiments to develop new space services. It can be expected that these lines of effort will create a cascade of counters and security dilemmas checked primarily by limited resources and competing problems.

Key Drivers of Change

While I attempted to paint a basic cartoon sketch of space warfare, there are a range of complexities which will give it shape. Below are trends I judge to have significant impacts on military space competition and conflict.

Increasing Nationalism: Nationalism is likely to be the single most significant factor accelerating military space capabilities. Since the beginnings of the first Space Race,[114] space has been the most public of arenas to showcase national strength, pride and vigor to internal and external audiences. Like a bird's plumage, it is a proxy for the strength and status of a nation. With the current trend in deglobalization,[115] return of great power competition,[116] nations are

---

spacecraft.

114. "The Space Race," *History.com*, February 21, 2020, accessed June 30, 2020, https://www.history.com/topics/cold-war/space-race.

115. Andrew Browne, "How the Coronavirus is Accelerating Deglobalization," *Bloomberg*, February 29, 2020, accessed July 1, 2020, https://www.bloomberg.com/news/newsletters/2020-02-29/why-deglobalization-is-accelerating-bloomberg-new-economy.

116. Says the US National Security Strategy, "after being dismissed as a phenomenon of an earlier century, great power competition returned. China and Russia began to reassert their influence regionally and globally"; the National Defense Strategy, "The central challenge to U.S. prosperity and security is the reemergence of long-term, strategic competition by what the National Security Strategy classifies as revisionist powers. It is increasingly clear that China and Russia want to shape a world consistent with their authoritarian model—gaining veto authority over other nations' economic, diplomatic, and security decisions"; the National Military Strategy, "These trends, especially those posed by the reemergence of great power competition with China and Russia, represent the

likely to measure each other based upon their relative positions in space. Lack of agreement over relative strength might create incentives for conflict.[117]

Increasing Militarism: With the rise of nationalism comes the desire to showcase relative strength, and to be able to respond to threats to interest or slights to status with posture, saber-rattling or decisive action. That in turn leads to incentives to arms race.

Increasing Environmental Concerns: There is an ongoing attempt to frame space issues as environmental concerns,[118] and to limit economic development[119] or weaponization based upon concerns for the space environment.[120] Significant events such as debris-causing

---

most difficult challenges facing the Joint Force" see: *National Security Strategy of the United States of America*, The White House, December 2017, accessed July 1, 2020, https://www.whitehouse.gov/wp-content/uploads/2017/12/NSS-Final-12-18-2017-0905.pdf ; Summary of the 2018 National Defense Strategy of the United States, Department of Defense, 2018, https://dod.defense.gov/Portals/1/Documents/pubs/2018-National-Defense-Strategy-Summary.pdf ; *Description of the National Military Strategy 2018*, Joint Staff, 2018, https://www.jcs.mil/Portals/36/Documents/Publications/UNCLASS_2018_National_Military_Strategy_Description.pdf.

117. James D. Fearon, "Rationalist explanations for war." *International organization* 49, no. 3 (1995): 379-414.

118. Rebecca Reesman, Michael P Gleason, Colleen Stover, and Layla Bryant, *Slash the Trash: Incentivizing Deorbit*, Aerospace Corp, April 2020, accessed June 29, 2020, https://aerospace.org/sites/default/files/2020-04/Reesman_SlashTheTrash_20200422.pdf.

119. Susanna Kohler, "Astronomy Impacts of Satellite Megaconstellations," *AAS NOVA*, April 13, 2020, accessed June 30, 2020, https://aasnova.org/2020/04/13/astronomy-impacts-of-satellite-megaconstellations/.

120. Secure World Foundation, Space Sustainability: A Practical Guide, 2018, accessed June 29, 2020, https://swfound.org/media/206407/swf_space_sustainability_booklet_2018_web.pdf ; Joan Johnson-Freese· *Heavenly Ambitions: America's Quest to Dominate Space* (Philadelphia, PA: University of Pennsylvania Press, 2009); James Clay Moltz, *Asia's Space Race: National Motivations, Regional Rivalries and International Risk* (New York: Columbia University Press, 2011).

tests[121] or collisions[122] might create disincentives for certain types of space warfare.

Increasing Volume of Activity: Two revolutions – low-cost reusable commercial launch and proliferation of large-production-run space hardware – provide the possibility of much larger scales of space activity, and new commercial locations and markets.[123] Larger scale of activity means much more frequent access to space[124] and within-space by logistical vehicles which in turn enable still larger markets.

Expansion of Information Services: Expanding global demand for bandwidth and computation will seek solutions anywhere it can sustain growth. One possibility is that this will drive new global utilities. Most likely are expansions of already extant capabilities: daily commercial Earth imaging,[125] commercial synthetic ap-

---

121. Brian Weeden and Victoria Samson, "India's ASAT test is wake-up call for norms of behavior in space," *Space News*, April 8, 2019, accessed June 29, 2020, https://spacenews.com/op-ed-indias-asat-test-is-wake-up-call-for-norms-of-behavior-in-space/ ; Leonard David, China's Anti-Satellite Test: Worrisome Debris Cloud Circles Earth *Space.com*, February 02, 2007, accessed June 29, 2020, https://www.space.com/3415-china-anti-satellite-test-worrisome-debris-cloud-circles-earth.html
122. By Loren Grush, "As Satellite Constellations Grow Larger, Nasa Is Worried About Orbital Debris: The space agency argues that these probes need to be de-orbited — reliably," *The Verge*, Sep 28, 2018, accessed June 29, 2020, https://www.theverge.com/2018/9/28/17906158/nasa-spacex-oneweb-satellite-large-constellations-orbital-debris.
123. Ivan Bekey, "Chapter 9: The Long-term Outlook for Commercial Space," in Charles D. Lutes, and Peter Hays, eds, *Toward a Theory of Spacepower: Selected Essays*, Figure 9–12. Potential Number of Annual World Launches, p. 196.
124. Casey Handmer, "The SpaceX Starship is a very big deal," *Casey Handmer Blog*, October 29, 2019, accessed June 28, 2020, https://caseyhandmer.wordpress.com/2019/10/29/the-spacex-starship-is-a-very-big-deal/#:~:text=Starship%20is%20not%20just%20a,This%20is%20a%20big%20deal.
125. "Planet Labs Imagery: The Entire Earth, Everyday," *GIS Geography*, April 26, 2020, accessed July 1, 2020, https://gisgeography.com/planet-labs-imagery/.

erture radar,[126] and commercial radio-frequency signal tracking.[127] Next likely is the extension of space services to un-modified cell phones,[128] and broad-band space-based internet.[129] We could also expect space-based full motion video, real-time space-based radar for traffic control for air, land and sea. It might also prove advantageous to host data-centers and computation centers in space.[130]

Increased Perceived Value of Space Resources: A key unknown is the speed at which space resources (lunar and asteroid mining) become important to national economies. Of note, space resources do not need to actually be productive to influence strategic behavior—the mere perception that they will be important is enough to initiate a logic of moves and counter-moves to secure or deny advantage.[131]

---

126. Rachel Jewett, "Capella Space Becomes First Commercial SAR Provider to Work With NRO," *Satellite Today*, December 12, 2019, accessed July 1, 2020, https://www.satellitetoday.com/government-military/2019/12/12/capella-space-becomes-first-commercial-sar-provider-to-work-with-nro/.
127. Annamarie Nyirady, "HawkEye 360 Releases 1st RF Signal Mapping Product," *Satellite Today*, April 4, 2019, accessed July 1, 2020, https://www.satellitetoday.com/imagery-and-sensing/2019/04/04/hawkeye-360-releases-1st-rf-signal-mapping-product/; Cortney Weinbaum, Steven Berner, and Bruce McClintock , "SIGINT for Anyone: The Growing Availability of Signals Intelligence in the Public Domain," RAND, 2017, accessed July 1, 2020, https://www.rand.org/content/dam/rand/pubs/perspectives/PE200/PE273/RAND_PE273.pdf.
128. Loren Grush, "Space startup Lynk uses satellite to send text message to unmodified Android phone: The company says it paves the way for its mega-constellation of satellite 'cell towers'," *The Verge*, March 18, 2020, accessed July 1, 2020, https://www.theverge.com/2020/3/18/21184126/lynk-mega-constellation-text-message-android-smartphone-cell-towers-space.
129. Casey Handmer, "Starlink is a very big deal," *Casey Handmer Blog*, November 2, 2019, accessed June 28, 2020, https://caseyhandmer.wordpress.com/2019/11/02/starlink-is-a-very-big-deal/.
130. Andrew Donoghue, "The Idea of Data Centers in Space Just Got a Little Less Crazy: Critical Thinking: SpaceX's latest launch raises the potential for all kinds of new space-based infrastructure, including data centers," *Data Center Knowledge*, February 09, 2018, accessed July 1, 2020, https://www.datacenterknowledge.com/edge-computing/idea-data-centers-space-just-got-little-less-crazy.
131. The Thomas Theorem States, "if men define situations as real, they are real in their consequences." Robert K. Merton., "The Thomas Theorem and The Mat-

Increased Space Access: The scope and scale of conflict depend greatly on the cost and reach of space access. Lower launch costs enable greater scale of activity[132]—higher launch tempos, more assets on orbit or on planetary bodies. Lower cost in-space transportation allows greater mobility, logistics, force extension and spatial extension of economic activity. The current trend suggests a lowering in the cost of space access to enable greater levels of economic and military activity.

Key Variables

In general, we can be confident that all of the above variables are increasing, the scale of change matters for the world we can expect in 1-2 decades. The range of such variables is discussed below.

Domain Ideology: A key variable is the understanding among the would-be belligerents about the place of space and the aims and limits of space warfare, as well as their differences. Different visions of space warfare and its ends drive different investment and posture strategies. Prescriptions for brown-water and blue-water acquisitions, organization, force structure, force posture, force employment and training diverge.

Economic Importance of Space: Space is already important and integral to the US and global economy, but the space sector itself is about $400B,[133] or less than half a percent of global GDP. At present, there is an expectation of an expanding cislunar econosphere[134]

---

thew Effect," *Social Forces*, 74/2 (1995): 379-424, available at http://garfield.library.upenn.edu/merton/thomastheorem.pdf.

132. Ivan Bekey, "Chapter 9: The Long-term Outlook for Commercial Space," in Charles D. Lutes, and Peter Hays, eds, *Toward a Theory of Spacepower: Selected Essays*, Figure 9–12. Potential Number of Annual World Launches, p. 196.

133. Timothy Taylor, "How Big is the Space Economy," *The Conversable Economist*, December 19, 2019, accessed June 28, 2020, https://conversableeconomist.blogspot.com/2019/12/how-big-is-space-economy.html.

134. Ken Murphy, "The cislunar econosphere (part 1)," *The Space Review*,

with a value in this time epoch to be between $1T[135] and $10T. [136] The economic importance of space might continue to rest (as it does today) on space information services. Alternately, 1-2 decades is enough to see the emergence of markets for space-based power[137]

February 20, 2012, accessed June 28, 2020, https://www.thespacereview.com/article/2027/1.

135. There is broad consensus that the space economy is poised to grow significantly. Major banks have provided forecasts circa 2040 (Goldman Sachs $1.1T Morgan Stanley $1.1T, U.S. Chamber of Commerce $1.5T) and circa 2050 (Bank of America Merrill Lynch $2.7T). Reflecting this optimism, the U.S. Secretary of Commerce gave a speech at the World Economic Forum in Davos where he said, "*Current industry projections place the 2040 global space economy at between $1 and $3 trillion.* And I think we will certainly get to a trillion before 2030." See: "Space: Investing in the Final Frontier," *Morgan Stanley*, November 7, 2018, accessed January 30, 2019, https://www.morganstanley.com/ideas/investing-in-space; Rich Smith, "The $1.1 Trillion Space Industry Prediction You Can't Afford to Miss," *The Motley Fool*, February 16, 2018, accessed https://www.fool.com/investing/2018/02/16/the-11-trillion-space-industry-prediction-you-cant.aspx; Brian Higgenbotham, "The Space Economy: An Industry Takes Off," U.S. Chamber of Commerce, November 11, 2018, accessed January 30, 2019, at https://www.uschamber.com/series/above-the-fold/the-space-economy-industry-takes; Michael Sheetz, "The space industry will be worth nearly $3 trillion in 30 years, Bank of America predicts," *CNBC*, October 31, 2017, accessed January 30, 2019, https://www.cnbc.com/2017/10/31/the-space-industry-will-be-worth-nearly-3-trillion-in-30-years-bank-of-america-predicts.html; Wilbur Ross, "A New Space Race: Getting to the Trillion-Dollar Space Economy World Economic Forum, Davos, Switzerland," Commerce Department, January 24, 2020, accessed February 21, 2020, https://www.legistorm.com/stormfeed/view_rss/1722262/organization/33279.html.

136. See discussions of China's "Cislunar Economic Zone (dì yuè kōngjiān jīngjì qū), or Moon-based economic sector" expected to return $10 trillion in annual economic benefit for China by 2050: Editors, "A Chinese Economic Zone In Space", *Sup China*, November 4, 2019, March 3, 2020, https://supchina.com/2019/11/04/a-chinese-economic-zone-in-space/; Andrew Jones, "From a far side first to cislunar dominance? China appears to want to establish 'space economic zone' worth trillions," *Space News*, February 15, 2020, accessed March 3, 2020, https://spacenews.com/from-a-farside-first-to-cislunar-dominance-china-appears-to-want-to-establish-space-economic-zone-worth-trillions/.

137. Research and Markets, "Global Space-Based Solar Power Market Study, 2020: An Emerging Industry with Unlimited Potential," *PR Newswire*, Feb 14, 2020, accessed July 1, 2020, https://www.prnewswire.com/news-releases/

and space-based resources,[138] each undergoing rapid growth.

Technological Progress in Access: NASA estimates that commercial launch lowered the cost of access by a factor of 20 vs the space shuttle ($54,500/kg vs $2,720/kg Falcon 9).[139] Falcon 9 recently demonstrated the feasibility of human-rated commercial launch. Falcon Heavy is now $1411 per kilogram[140] (a factor of 38 better than the Space Shuttle). It is possible that launch costs could remain above $1000/kg for this epoch. Some allege that this is already low enough to enable space-based solar power. On the other end, SpaceX Starship with its combination of full reusability and 100MT+ payload might enable a reduction of cost to LEO two orders of magnitude to $10/kg and perhaps $30/kg to the lunar surface.[141] It has been previously estimated that emplacing as little as 12MT on the lunar surface is sufficient to begin industrial 'bootstrapping' and exponential increase in industrial activity.[142]

---

global-space-based-solar-power-market-study-2020-an-emerging-industry-with-unlimited-potential-301005179.html.

138. Leonard David, "The US Geological Survey Is Getting Serious About Space Resources and Mining," *Space.com*, September 04, 2018, accessed July 1, 2020, https://www.space.com/41707-space-mining-usgs-resource-survey.html.

139. Harry W. Jones, "The Recent Large Reduction in Space Launch Cost," NASA, 8-12 July 2018, accessed June 22, 2020, https://ntrs.nasa.gov/archive/nasa/casi.ntrs.nasa.gov/20200001093.pdf.

140. Brian Wang, "Reusable SpaceX Falcon Heavy will make highly cost competitive space based solar power", *Next Big Future*, February 5, 2018, accessed June 22, 2020, https://www.nextbigfuture.com/2018/02/reusable-spacex-falcon-heavy-will-make-highly-cost-competitive-space-based-solar-power.html#:~:text=A%20SpaceX%20Falcon%20Heavy%20will,will%20be%20%241411%20per%20kilogram.

141. Ramish Zafar, "SpaceX Could Bring Starship Launch Costs Down To $10/kg Believes Musk," *WCCFTech*, May 8, 2020, accessed June 22, 2020, https://wccftech.com/spacex-launch-costs-down-musk/#:~:text=These%20are%20military%20payloads%2C%20with,is%20certified%20to%20deliver%20them.&text=Returning%20to%20the%20moon%20missions,kg%20for%20delivering%20lunar%20payloads.

142. Metzger, Philip T., Anthony Muscatello, Robert P. Mueller, and James Mantovani. "Affordable, rapid bootstrapping of the space industry and solar system civilization." *Journal of Aerospace Engineering* 26, no. 1 (2013): 18-29, accessed June 28, 2020, https://arxiv.org/abs/1612.03238.

Military Readiness: There appears to be a general trend toward the militarization of space likely with some level of anti-satellite weaponization. Investments in military space readiness will be a function of ideology, threat, available budget and competing priorities. Either belligerent might underprepare for the 'right' space war, or could miscalculate the actual theater and be prepared for the 'wrong' space war.

Implications for Great Power Competition: The Will to Prepare and Shape

This paper has attempted to leave the reader with two distinct images of warfare. The first presumes that a space war is an extension of Earthly terrestrial geo-strategic concerns and extends to space to preserve overhead advantage. The second presumes that states have geo-strategic concerns in the space domain itself which generate conflict. A strategist's or policy maker's assumptions about the reasons for war, the object of such a war, strongly affect one's planning, preparation and force structure. A focus on one to the exclusion of the other may result in incomplete preparation.

Theory Shapes Reality -- States Will Prepare along Lines of Expectation: The Thomas theorem states, "if men define situations as real, they are real in their consequences."[143] Visions of astrostrategic competition and what is considered 'strategic' – even if it proves not to be -- will shape state action. Visions of space warfare – even if it does not happen – will shape state action. The contours of competition are shaped by what is defined as inherently valuable within space, and what is defined as instrumentally valuable in space to secure, protect, defend, or attack. Almost all 'inherently valuable' objects are themselves instruments as part of broader means-ends chains. Today, satellites in LEO and GEO can be said to be inherently valuable because they provide valuable services economically

143. Robert K. Merton., "The Thomas Theorem and The Matthew Effect," *Social Forces* 74, no. 2 (1995): 379-424, available at http://garfield.library.upenn.edu/merton/thomastheorem.pdf.

and militarily on Earth. Space control assets are instrumentally valuable because they can secure or deny such value. If, on the other hand, one focuses on the ice deposits on the poles of the Moon and certain asteroids as inherently important resources and certain locations when enable their exploitation (key landing locations, well-illuminated crater rims), the Earth-Moon Lagrange points, you shape your strategy around to accessing and securing these. In short, you end up with a very different military space architecture and set of military and commercial strategies.

Astrostrategic Terrain: The main new competition will be to enable or avoid coercive advantage in the astrostrategic terrain beyond GEO to enable freedom of action and acquiescence to one's will. This is a competitive posture game likely to escalate from simple presence to space information services, space domain awareness, dual-use logistics systems, self-defenses, and perhaps 'snugglers,' soft-mission-kill capabilities (jamming, dazzling) and even long-range direct-fire systems.

Dynamics Incentivizing Competition: Several additional factors seem likely to accelerate military competition in space. These include commercial actors militating for the 'flag to follow trade' and provide security services for them, as well as provocative action on their part to embroil their state in activities which create private benefit. Such provocative action could be self-help or the threat of self-help should they not receive government security services. A second is a classic security dilemma where the actions by one party to make itself more secure make other parties less secure. Efforts to deter the other by increasing the cost of intervention may just as easily result in greater military investments. A third is the lack of a clear military order in space which parties may seek to solve through investment or resolve through battle.[144] A fourth incentive are audience costs for reputation or strength which transfer to other theaters where perceived pre-eminence and power affects bargaining and access to spoils. A special case of the above may concern public or

144. James D. Fearon, "Rationalist explanations for war." *International organization* 49, no. 3 (1995): 379-414.

treaty commitments to third parties (such as an allied nation) whose geo-economic or personnel might be the subject of coercion.

Prospects for Arms Control or Restraint in Arms Conflict

Prospects for restraint appear more likely if there is an early conditioning shock which results in costs with few benefits, horror, or naturally conditions cooperation and good will. Examples of the former might include a limited ASAT exchange by spacefaring states (perhaps India-Pakistan or even India-China) which shocks the major powers, or an attempt by a rouge state, polity or lone actor to hold the world hostage through deliberate use of a collision, space debris or an Electro Magnetic Pulse (EMP) delivered by a High Altitude Nuclear Detonation (HAND) (perhaps by Iran or North Korea).[145] A successfully negotiated crisis de-escalation is likely to result in a series of 'Détente'[146] measures to live together and reduce provocations to conflict as well as agreements on 'dangerous practices at sea.' In all cases, such a direct encounter with a general war in space could reduce the appetite for competitive and escalation under the sentiment 'too rich for me!'

Arms control in expansionist scenarios may be more likely if both parties see themselves in a precarious position both logistically and with their domestic audience. This might encourage mutually respected safety zones and protocols for safe interaction. Early efforts to limit a cycle of retaliations could focus on deliberate exercises to provide emergency aid in distress, and as much precision and clarity on boundary lines and coordinate systems as is possible. Further, while dual-use infrastructure may impact warfare in both conserva-

145. David Stuckenberg and Peter Garretson, "Updating the Laws of War for the 21st Century: Time to Reconsider, and Ban, Modern Nation-Killing Acts," The Mitchell Institute, April 2, 2019, accessed July 1, 2020, https://www.mitchelaerospacepower.org/single-post/2019/04/02/Updating-the-Laws-of-War-for-the-21st-Century-Time-to-Reconsider-and-Ban-Modern-Nation-Killing-Acts.
146. "Détente," *History.com*, October 27, 2009, accessed June 29, 2020, https://www.history.com/topics/cold-war/detente.

tive and expansionist scenarios, at the extremes of logistical reach, very tiny influences may enable exclusion, and dual-use infrastructure may present strong concerns. The inability to limit dual-use infrastructure because of its critical functions may force arms control to focus less on numerical limits and more on inspection, verification and multi-lateral controls.

Chapter 3:

Commercial Space: Space Controls and the Invisible Hand

Brian G Chow[147]

Abstract

This article reviewed three major projections of the global space industry by Goldman Sachs, Morgan Stanley, and Bank of America Merrill Lynch and extracted the trends that would significantly impact the design of both the domestic and international space traffic management (STM) schemes. It found that, in the next two decades, the United States will still have the largest market share in practically every space industrial sector. It suggests how the United States, as well as the West, can use its market power to incentivize Russia and China to fall in line with a STM that provides peace and prosperity to all. It also proposed five measures as building blocks for developing standards, practices, regulations and laws for such a STM.

This article explores how by the use of space controls and the invisible hand in commerce, space can continue to be a peaceful and prosperous place. Let's first explain what "commercial space", "space controls", and "invisible hand" mean.

- The National Space Policy issued on June 28, 2010 contained

147. I would like to thank the Nonproliferation Policy Education Center for providing financial support to the preparation of this research article.

the commercial space guidelines, which stated:

> "The term 'commercial,' for the purposes of this policy, refers to space goods, services, or activities provided by private sector enterprises that bear a reasonable portion of the investment risk and responsibility for the activity, operate in accordance with typical market-based incentives for controlling cost and optimizing return on investment"[148]

The customers for commercial space are from both the private and public sectors; and both domestic and foreign parties.

- This article argues that, for space control to be efficient, it should be applied through the norms, regulations and laws in commercial space and backed up by military recourse as a last resort.

- Adam Smith said that the economy can work well, if it is guided by the invisible hand, in which everyone works for his or her own interest. This article argues that this invisible hand can be used in commercial space to help keep space activities profitable and peaceful.

Section I describes the important role of space traffic management (STM) to enable the global space industry to provide both economic prosperity and military security, including the protection of our critical satellites against robotic and other antisatellite (ASAT) attacks. From three major reports about the future of the global space industry, Section II extracts the key trends that will have significant ramifications on space threats and controls. Section III identifies measures that can help avoid business ruin (i.e. economic security) and buy warning time to protect satellites (i.e. military security). These measures can be used as the building blocks for developing standards, practices, regulations and laws for governing domestic and international space traffic. Section IV provides the conclusion and summarizes the recommended measures.

---

148. National Space Policy of the United States of America, June 28, 2010, https://history.nasa.gov/national_space_policy_6-28-10.pdf.

I. Space Traffic Management Is the Foundation of Prosperous and Peaceful Space

The Space Policy Directive-3, National Space Traffic Management Policy (Directive-3) unveiled by President Trump on June 18, 2018 states that "[t]o maintain US leadership in space, we must develop a new approach to space traffic management (STM) that addresses current and future operational risks."[149] Indeed, there is only one single outer space and one single STM for all spacefaring nations of diverse interests to operate in. This article believes that a principled and equitable STM holds the key to prosperous and peaceful space for all.

The Directive-3 also states:

> "As the leader in space, the United States supports the development of operational standards and best practices to promote safe and responsible behavior in space. A critical first step in carrying out that goal is to develop US-led minimum safety standards and best practices to coordinate space traffic. US regulatory agencies should, as appropriate, adopt these standards and best practices in domestic regulatory frameworks and use them to inform and help shape international consensus practices and standards."[150]

This is a practical step-by-step plan to help establish a global STM policy by proactively participating in the development of standards and practices; following up with regulations; and demonstrating how well the STM works for US systems in order to help shape the international STM.

However, even this good action plan would not eliminate disagreements among stakeholders, such as commercial space equipment

---

149. The White House, Space Policy Directive-3, National Space Traffic Management Policy, Presidential Memoranda, June 18, 2018, https://www.whitehouse.gov/presidential-actions/space-policy-directive-3-national-space-traffic-management-policy/.
150. Ibid.

and service providers and government users, because naturally they have different views of how the development of a STM system should proceed and how the STM should work to protect their interests. For example, the Commercial Spaceflight Federation, a Washington-based industry association of leading businesses in the field, has been reported to be pushing for a voluntary consensus approach to safety standards, while Tommaso Sgobba, the Executive Director of the nonprofit International Association for the Advancement of Space Safety and the former head of the European Space Agency's Independent Safety Office, argued that "[t]here is in the world no safety-critical system which is certified on the basis of a voluntary standard. In the case of a standard, there always needs to be some of enforcement."[151] Also, Theresa Hitchens reported that "[a] growing number of US satellite owners/operators want to see new on-orbit safety rules" but "[n]onetheless, industry officials say, there remain strong pockets of resistance to any new regulations especially among aggressive space startups,"[152] as new regulations can add cost or complicate operations.

On the other hand, while disagreements complicate the process, they do not prevent the establishment of a good STM so long as the STM is fair to stakeholders and meets their key requirements in, such as, profitability and safety, as they realize that each of them must make some sacrifice for the common good.

It is easier for other western countries coming to a compromise with us in the STM principles and rules than our potential adversaries, who could prefer the status quo where they can threaten our satellites. Therefore, our strategy needs to pursue a multinational STM and an international STM at the same time. For the former, we focus on getting an agreement with the West[153] and whoever else wants to

151. Tereza Pultarova, Independent body proposed to ensure commercial spaceflight safety, *SpaceNews*, April 26, 2019, https://spacenews.com/independent-body-proposed-to-ensure-commercial-spaceflight-safety/.
152. Theresa Hitchens, Satellite Firms Debate New Commerce Space Safety Rules, *Breaking Defense*, April 26, 2019, https://breakingdefense.com/2019/04/satellite-firms-debate-new-commerce-space-safety-rules/.
153. The article takes a broad definition for the West (i.e. western countries),

join to show the structure of a proper STM. For example, as a condition of doing business with the United States and other western countries, we establish a rule that does not allow satellites, domestic or foreign, to come within a minimum safe distance from our critical satellites, unless exempted by prior consent. If China and Russia insist not to observe this rule so as to threaten our critical satellites at time of their choosing, they simply cannot do business with us. Once potential adversaries recognize that the West is pursuing a fair (including to our adversaries) STM, they may well reevaluate their own positions and find themselves better off by joining the realistic approach of the United States and other nations in keeping the peace and prosperity in space.[154]

II. Impacts of the Future Commercial Space on Space Threats and Controls

This section extracts the key trends from three major reports by Goldman Sachs,[155] Morgan Stanley,[156] and Bank of America Merrill Lynch (BofAML)[157]. These key trends will have significant ramifications on how the robotic threat will rapidly grow and how commercial space activities can contribute to not only economic prosperity but also military security such as protecting our satellites against

which includes NATO member countries; Australia, New Zealand, Japan, South Korea, Taiwan, Israel; and the non-aligned Austria, Finland, Sweden, and Switzerland. (See Western countries, fact-index.com, http://www.fact-index.com/w/we/western_countries.html.)

154. This article will return to this approach for getting Russia and China to fall in line with the West in Section V.

155. Goldman Sachs, Space: The Next Investment Frontier, April 4, 2017, http://www.fullertreacymoney.com/general/space-the-next-investment-frontier-/.

156. Morgan Stanley, Space: Investment Implications of the Final Frontier, October 12, 2017, http://www.fullertreacymoney.com/system/data/files/PDFs/2017/October/20th/msspace.pdf.

157. Bank of America Merrill Lynch, To Infinity And Beyond – Global Space Primer, October 30, 2017, https://go.guidants.com/q/db/a2/1e1ffc185c1d44bd.pdf.

ASAT threats including those from robotic spacecraft[158].

A. Key Space Operators in Various Commercial Space Sectors

Figure 3.01 shows the major spacefaring nations' government spending in space. The United States accounts for about 50% of the space spending throughout 2018 to 2040. Including other western countries shown in Figure 3.01, it increases to 75%. This clearly shows the powerful influence of US in particular and the West in general. Also, as space will soon be weaponized, the no-space-weapon approach long led by Russia and China is bound to lose favor. The West should offer and demonstrate an alternative approach for the newly weaponized space in their space policy and domestic STM, in addition to working with Russia, China and other countries to shape an international STM for all. This two-pronged approach is similar to the approach of demonstrating domestically and shaping international STM in Directive-3. Once the western countries are unified to offer the same standards, regulations and laws for domestic and foreign commercial space companies that do business with them, the space companies would want to simplify their operations by pushing for the non-western countries such as China and Russia to adopt the same practices. This approach will help shape and quicken the establishment of the international STM.

158. Spacecraft and satellite are used interchangeably. However, when a satellite is designed to perform rendezvous and proximity operations frequently, it tends to be called a spacecraft.

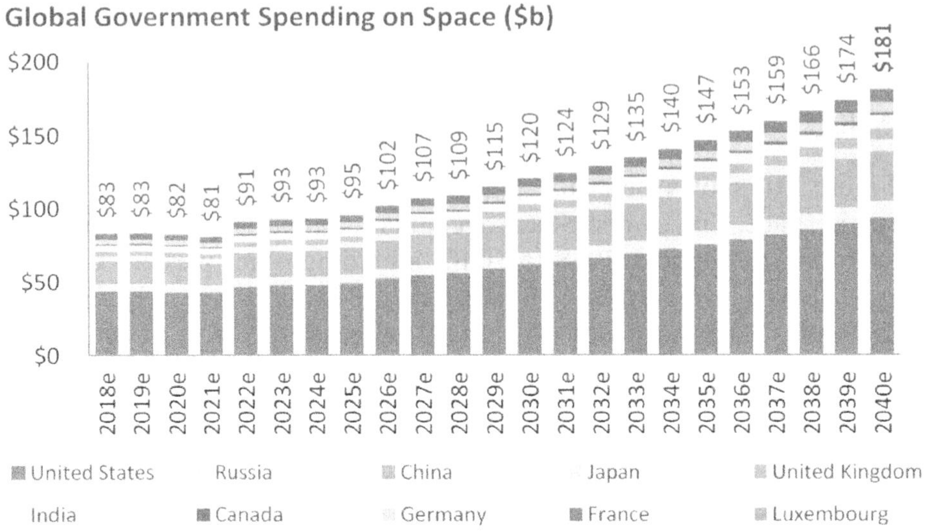

Source: Morgan Stanley, *op. cit.*, p. 29.

Figure 3.01. Western Countries Shown Account for ¾ of Government Space Spending.

Figure 3.02 exhibits the share of satellites by country. It shows a similar dominance of western countries over Russia and China, which together account for merely 23% of the number of operating satellites.

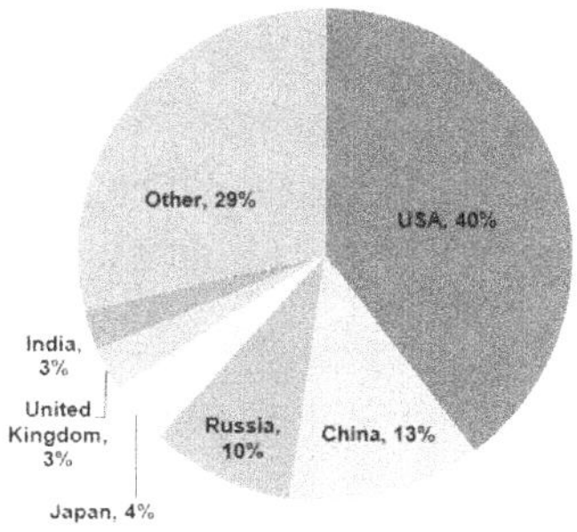

Source: Goldman Sachs (derived from Union of Concerned Scientists), *op. cit.*, p. 13

Figure 3.02. Share of Satellites by Country.

Thus, the United States has the largest share of operating satellites and government spending and the West has a total share much larger than those of China and Russia. The western countries should unify their space policies and shape the international STM.

Figure 3.03 shows the global space economy breakdown in 2016. The global space economy reached \$339 billion. Satellite Services in telecommunications, earth observation, science and national security accounted for \$127.7 billion; ground equipment such as satellite TV, radio and broadband (e.g. high-speed internet access) equipment, and Global Navigation Satellite Systems and chipsets, \$113.4 billion; satellite manufacturing, \$13.9 billion; and the launch industry, \$5.5 billion. The satellite industry, which is commonly defined as consisting of these four segments, accounted for \$261 billion or 77% of the global space economy (i.e. industry), while the non-satellite industry accounted for the remaining 23% or \$78.6 billion.

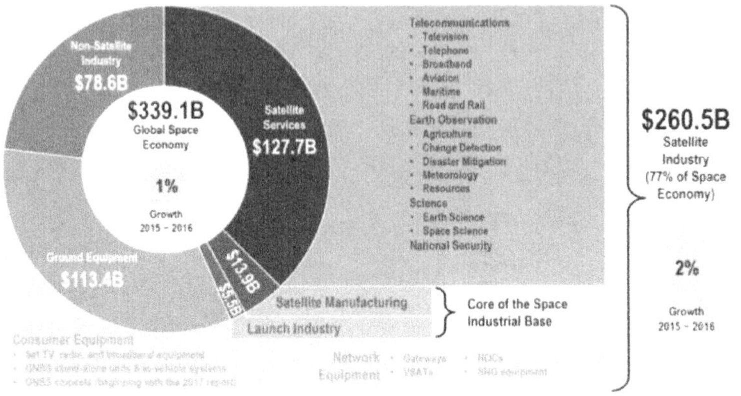

Source: Bank of America Merrill Lynch (derived from SIA, Tauri Group, Bryce Tech), *op. cit.*, p. 35.

Figure 3.03. Global Space Market Breakdown in 2016.

1. Launch Industry

Let's start with the launch industry. While this is the smallest sector of the satellite industry and has merely a 2% share, it has the most exciting innovation and shows what the United States can do to revolutionize the launch sector, as well as the entire global space economy.

Table 3.01 shows a list of key launch providers. Of the fifteen launch providers, the United States accounts for eleven of them.

Country	Launch Provider
USA	Blue Origin
USA	Boeing
USA	International Launch Services
USA	Lockheed Martin
USA	Orbital ATK (Northrop Grumman)
USA	Rocket Lab
USA	SpaceX
USA	Stratolaunch
USA	Vector
USA	Virgin Galactic
USA	XCOR Aerospace
France	Safran (owns 50% ArianeGroup)
Nether-lands	Airbus (owns 50% ArianeGroup)
Japan	Mitsubishi Heavy Industries
China	China Great Wall Industry Corp.

Source: Goldman Sachs, *op. cit.*, p. 8.

Table 3.01. Key Launch Providers Worldwide.

Goldman Sachs reported that the United States share of the global launch revenues during 2006-2016 was 19%.[159] Then, it jumped to 47%[160] largely because SpaceX quickly captured a large number of launches, as it played the key role in reducing the launch cost by a large factor.[161] By 2040, Morgan Stanley projected SpaceX to have about 60% of the global launch market. [162] At worst, it would probably form a duopoly with Arianespace, a subsidiary of ArianeGroup. The combined launch share of SpaceX and Arianespace will continue to be far higher than that of Russia and China. SpaceX's achievement in space launch is due to its pathbreaking innovation and the willingness to risk huge sum of capital expenditure. SpaceX's achievements also force both longtime and start-up competitors to strive for lower price and/or better quality. On the other hand, Chris Boshuizen, entrepreneur in residence at venture capital firm Data Collective pointed out that there are more than 100 small launch vehicles in various stages of development.[163] The number would increase, if launch vehicles in heavier classes are included. It is clear that the launch market is highly competitive and overcrowded. A few will likely capture the dominant share of the launch market, with many more will fall by the wayside.

Goldman Sachs reported that the Chinese Long March rockets are priced very competitively and have low failure rates, but the United States and European launch providers are not worried because there are regulatory barriers that prevent US components from flying on

159. Goldman Sachs, *op. cit.*, p.3.
160. Ibid.
161. BofAML reported that "on a per kilogram of thrust basis launch costs have fallen from around US$10,000 per kg/LEO in 1967 when Saturn V launched to around US$2,600 per kg/LEO in 2016 with the Falcon 9 v1.2 (Full Thrust) (source: FAA). This could fall even further with Falcon Heavy, which is due for testing from late 2017 with a cost of only c.US$1,400 per kg/LEO, implying launch costs will have fallen by a factor of 10x with its introduction. Elon Musk believes that 'when upper/second stage & fairing are also reusable launch costs will drop by a factor of more than 100x'" (See BofAML, *op. cit.*, p. 32).
162. Morgan Stanley, *op. cit.*, p. 25.
163. Jeff Foust, Investors seek disruptive space startups in an overcrowded market, *SpaceNews*, May 7, 2019, https://spacenews.com/investors-seek-disruptive-space-startups-in-an-overcrowded-market/.

Chinese rockets.[164] Nearly all European satellites contain US components.[165] So, Chinese rockets cannot be used to launch US satellites as well as most of the European satellites.

Among the four space industrial sectors,[166] Russia has long been most prominent in space launches, but yet the whole launch sector accounts for merely 2% of the satellite industry. The US-based International Launch Services markets Russian Proton and the new Angara[167] launchers. The company has conducted merely five commercial Proton launches since the end of 2015 with the most recent mission in September 2017.[168] For almost two decades, the United States has relied on the Russian RD-180 engine in Atlas 5 to power national security space launches. Yet, Lara Seligman reported that "[t]he Defense Department, is racing toward a congressionally mandated deadline of December 2022 to fly the first all-American rocket, powered by domestically produced engines, for US national security space launches."[169] Worst yet, Russian Deputy Prime Minister Dmitry Rogozin believes that it is not worth the effort for Russia to try to elbow SpaceX and China aside in the market for launch vehicles.[170] His statement signals that Russia is thinking to giving up competing in the market to launch western satellites.

---

164. Goldman Sachs, *op. cit.*, p. 29.
165. Ibid. See also Peter B de Selding, European Satellites Still Heavily Dependent on U.S. Parts, *SpaceNews*, January 29, 2015, https://spacenews.com/european-satellite-still-heavily-dependent-on-u-s-parts/.
166. In 2016, Satellite services sector accounts for $127.7 billion; ground equipment sector, $113.4 billion; satellite manufacturing sector, $13.9 billion; and launch sector, $5.5 billion or 2% (i.e. 5.5/261).
167. Angara-5 to replace Proton, Russian Space Web, downloaded on June 12, 2019 from http://www.russianspaceweb.com/angara5.html.
168. Stephen Clark, Proton launch provider ILS embraces closer relationship with roscosmos, Spaceflight Now, May 17, 2019, https://spaceflightnow.com/2019/05/17/proton-launch-provider-ils-embraces-closer-relationship-with-roscosmos/.
169. Lara Seligman, The New Space, Foreign Policy, May 14, 2019, https://foreignpolicy.com/2019/05/14/the-new-space-race-china-russia-nasa/.
170. Russian deputy PM sees no reason for competing with Musk on launch vehicles market, TASS, April 17, 2018, http://tass.com/science/1000229.

Thus, both China and Russia are weak in competing with the West for launching satellites, except those from their own countries. The defense market is fragmented, because countries including China and Russia provide national support to launch providers from their own countries as shown in Figure 3.04.

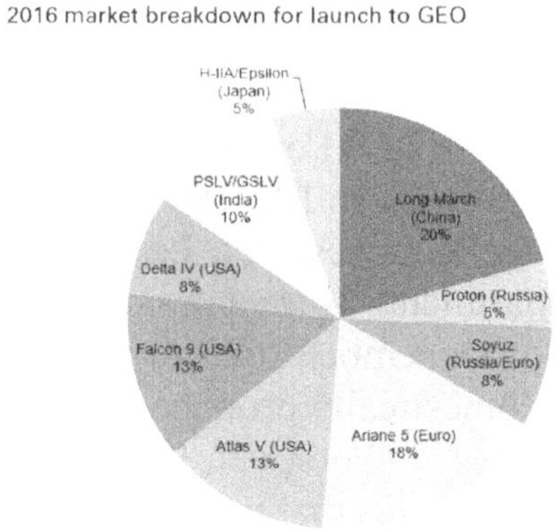

Source: Goldman Sachs, *op. cit.*, p. 24.

Figure 3.04. Countries Tend to Support Domestic Providers for Defense Launches.

In sum, the United States in particular and the West in general will continue to dominate the launch market.

2. Satellite Manufacturers

Table 3.02 shows the key satellite manufacturers worldwide. Nine out of thirteen are US satellite manufacturers. Goldman Sachs reported that "[m]ost satellites are built in the US or Europe."[171] Goldman Sachs also said that "the nascent Chinese satellites manufactur-

171. Goldman Sachs, *op. cit.*, p. 13.

ers have yet to prove their technology over a meaningful period of time. Their oldest satellite is 6 years old according to the Union of Concerned Scientists database as of June 2016."[172] Thus, China has too little experience in satellite manufacturing to capture any sizable business from the West. Also, while Russia will try to get into satellite manufacturing,[173] it has little resources for catching up with, and wrestling any sizable share from, the West.

Country	Satellite Manufacturers
USA	Boeing
USA	Harris Corp
USA	L3 Technologies
USA	Lockheed Martin
USA	MDA
USA	Northrop Grumman
USA	Raytheon
USA	SpaceX
USA	ViaSat
France	Thales
Netherlands	Airbus
Brazil	Embraer
China	China Great Wall Industry Corp.

Source: Goldman Sachs, op. cit., p. 8.

Table 3.02. Key Satellite Manufacturers Worldwide.

172. Ibid., p. 29.
173. Brian Wang, China main SpaceX competitor as Russia is giving up, Next Big Future, May 14, 2018, https://www.nextbigfuture.com/2018/08/china-main-spacex-competitor-as-russia-is-giving-up.html.

3. Satellite Services Operators

Table 3.03 shows that nine of fifteen key satellite services operators are located in the United States, and the rest of the fifteen are in Europe. Again, companies from the United States and Europe dominates satellite services worldwide. However, when Goldman Sachs assembled the key satellite services operators in 2017, it missed SpaceX's subsequently rapid rise in satellite services. As a result, this article delays the discussion of the future of Russia and China in satellite services until Section II.B where SpaceX's plan for satellite services is discussed.

Country	Satellite Manufacturers
USA	AT&T
USA	Deep Space Industries
USA	Digital Globe
USA	Dish Network
USA	Iridium
USA	Planet Labs
USA	Planetary Resources
USA	Spire
USA	ViaSat
France	Eutelsat
Netherlands	Airbus
Luxembourg	Intelsat
Luxembourg	SES
UK	Inmarsat
UK	OneWeb

Source: Goldman Sachs, op. cit., p. 8.

Table 3.03. Key Satellite Services Operators Worldwide.

B. Size and Profitability of the Future Global Space Industry

Morgan Stanley reported that, in 2016, government purchases

accounted for $84 billion or 25% of the $339 billion[174] global space industry revenues. They also projected that the industry will grow to $1.1 trillion by 2040 in their base case.[175]

The tripling of the global space industry will make the commercial space operators have an even stronger influence on space policy, risking that the STM will be constructed to favor economic prosperity more than military security. Also, the market share of government users worldwide will drop from 25% to 17%. However, this reduced but still significant share (an absolute amount of $181 billion[176]) will allow governments worldwide to continue having strong market influence, in addition to their political and regulatory power, in shaping the STM both domestically and internationally.

That governments will remain influential is consistent with the assessment of BofAML, whose US Aerospace & Defense team believes:

> **"...that military interests will remain the fundamental driver for investing in space.** In addition to ongoing activity (e.g. proposed United States Space Corps and tests of anti-satellite weapons (ASAT)), recent M&A [merger & acquisition] activity means it is not hard to imagine a future in which space-based direct energy missile defense systems are developed. Ongoing geopolitical concerns could be another tailwind for the potential weaponization of space."[177]

BofAML also found space investments to be risky and challenging:

> "There remain significant risks and challenges to investing in space and several technical hurdles to overcome before 'lift off'. Beyond the capex [capital expenditure] requirements, technical risks to the new Space Race include: unprofitability (low margins/returns), failed launches (1/20

---

174. Morgan Stanley, op. cit., p. 11
175. Ibid., p. 26.
176. Ibid., p. 11.
177. BofAML, op. cit., p.3.

> chance), safety concerns and regulation. Outer space risks include space debris (170mn [million] pieces in orbit) and solar storms (up to US$2.6tn [trillion] in losses)."[178]

It also reported:

> "16 of the world's 500 richest people, with a net worth of over US$500bn, have space investments."[179]

But it concluded:

> "Whilst the investment out of pocket by these billionaires is a positive for [s]pace we stress to investors that ultimately does not change the physics of their endeavor as space remains a high risk are of risk/reward."[180]

SpaceX's ambition also extends to satellites for providing internet access to people around the world. By 2040, Morgan Stanley projects that the largest share of the $1.1 trillion revenues for the global space industry will go to internet services for $410 billion.[181] SpaceX is planning to spend $10-15 billion for its Starlink small satellites (smallsats) in low Earth orbits (LEOs) to provide global internet access. Since a single geosynchronous (GEO) satellite can stare at 42%[182] of the Earth at all times, it alone can provide services to a local area at all time. In contrast, Mark Juncosa, SpaceX's vice president of vehicle engineering, said 12 Starlink launches, each carrying roughly 60 LEO smallsats, would cover the United States; 24 launches, most of the world's population; and 30 launches, the planet.[183] Therefore, business economics favors LEO satellite constellations which serve the world. For this reason, the lion share of the satellite

---

178. Ibid., p.4.
179. Ibid., p. 34.
180. Ibid.
181. Morgan Stanley, op. cit., p. 5.
182. T.S. Kelso, Basics of the Geostationary Orbit, downloaded on May 15 from https://celestrak.com/columns/v04n07/.
183. Caleb Henry, Musk says Starlink "economically viable" with around 1,000 satellites, *SpaceNews*, May 15, 2019, https://spacenews.com/musk-says-starlink-economically-viable-with-around-1000-satellites/.

access to internet will likely go to only a few providers worldwide, who can afford the huge capital expenditure; provide good services at competitive price; and capture a big chunk of the market first. As the West has already captured a dominant share in both the demand and supply of the space products and services worldwide, the major western countries will strive to control the largest global space sector (i.e. the $410-billion internet services) for economic prosperity and military security. While Russia will also try to get into satellite services,[184] there is little chance that Russia and/or China can get a sizable share of the western satellite services or other space markets, because they are not proficient enough to compete in the western free market especially in the space industry, which demands the use of latest technologies and rapid and frequent innovations. The best they can do will likely be to enter into a partnership with the West in order to share the benefits from space. But then, they would have to follow the STM rules established by the West.

On the other hand, capturing a lion share of a space market segment by a small number of providers does not mean capturing 100%. For its own economic and military security, a major spacefaring country likely wants to own some of the launch and other space industrial capabilities so as not to totally rely on other countries to provide these critical space services. It would likely subsidize some of its indigenous space capabilities. Therefore, the best cost/quality providers in the world would not capture the whole market.[185]

C. Implication of Smallsats on ASATs

In 2017, Morgan Stanley reported that satellite broadband[186] startup OneWeb will launch 2,740 smallsats into LEOs and medium-Earth

184. Wang, *op. cit.*
185. See Figure 3.04 and its discussion.
186. Satellite broadband means a wide bandwidth that carries multiple channels at once and provide high-speed internet access to many users at the same time.

orbits (MEOs).[187] The latency[188] will be about 50 milliseconds (ms), while a legacy satellite such as ViaSat 1 at GEO[189] is rated at 600 ms. Other than the need to provide internet access to most of the world's population as indicated in Section II.B, low latency (e.g. 50 ms) provided by LEO satellites internet access is highly attractive for interactive services, such as two-way chat and gaming, because one party can only respond after the message from the other party or parties are received and nobody wants that time lag to be noticeable.[190]

OneWeb is planning to launch 600 first-generation smallsats by 2019 or 2020 for its initial LEO constellation. Earlier, it was planning for 900, but 600 is the minimum needed for a global coverage. It said that more than $2 billion has been raised to date.[191]

Morgan Stanley also reported that SpaceX planned to launch 11,943 smallsats to LEO in two phases: 2019 to 2024, and 2029-2032 for full capacity at a cost of $10 to $15 billion.[192] The latency would be 20 to 30 ms. SpaceX launched 60 of its Starlink broadband satellites on May 23, 2019, all on one of its Falcon 9 launchers, to LEOs.[193]

187. Morgan Stanley, op. cit., p. 21.
188. Latency is the roundtrip time for a signal "to travel from your home to the satellite in orbit above the Earth, and then down to a ground-based gateway which connects you to the internet." (See Alex Miller, Satellite internet latency: What's the big deal?, Viasat | exede, September 6, 2017, https://www.exede.com/blog/satellite-internet-latency-whats-big-deal/.
189. ViaSat 1 is "the highest capacity satellite in the world." It "entered commercial service on January 16, 2012." "In late 2013, the technology began delivering the fastest Wi-Fi in the sky to airline passengers aboard JetBlue Airways." (See ViaSat-1 Launch, Viasat, downloaded on May 14, 2019, https://www.viasat.com/viasat-1-launch). It has a latency of 600 ms, because it is located in geosynchronous orbit. It weighs 6,740 kg. (See ViaSat1, Gunter's Space Page, downloaded on May 14, 2019, https://space.skyrocket.de/doc_sdat/viasat-1.htm).
190. In contrast, ViaSat 1 at GEO with a high latency of 600 ms is good for broadcasting, which is basically one-way transmission.
191. Caleb Henry, OneWeb scales back baseline constellation by 300 satellites, December 13, 2018, https://spacenews.com/oneweb-scales-back-constellation-by-300-satellites/.
192. Morgan Stanley, *op. cit.*, p. 22.
193. Caleb Henry, SpaceX launches 60 Starlink satellites, begins constellation buildout, *SpaceNews*, May 23, 2019, https://spacenews.com/spacex-launches-

These satellites will be followed by thousands more. As of March 31, 2019, Union of Concerned Scientists estimated that there are 2,062 operating satellites worldwide with 1,338 at LEOs.[194] Just SpaceX alone will raise the number of LEO satellites worldwide by a factor of nine by 2032.

These numerous smallsats may stress the ability to assess space situational awareness. The smallsats may need to be equipped with identification beacons so that their whereabouts can be easily detected to avoid accidental, as well as intentional, collisions. Two researchers at the Aerospace Corporation said that "[t]he goal is to design an inexpensive, small, and self-sufficient hosted payload package that will report data at regular intervals to enhance space situational awareness and better enable space traffic management activities."[195] They listed five potential payload packages including the 9x6x1.5cm GPS Transponder invented at their company and the 2x2x0.5cm Extremely Low Resource Optical Identifier[196] invented at the Los Alamos National Laboratory.

However, what if our adversaries design some smallsats for an ASAT purpose? Already, "[r]umors have been circulating for years that the Chinese Communist Party (CCP) has developed small satellites with robotic arms that could be used as anti-satellite weapons" and "[S]ome of the smaller satellites are lighter than 22 pounds, yet have a triple-eye sensor to gauge the shapes of targets and can adjust their speed and rotation, allowing them to grab objects within a distance of six inches, using a single robotic arm."[197] Although this

---

60-starlink-satellites-begins-constellation-buildout/.

194. Union of Concerned Scientists, UCS Satellite Database, downloaded on June 12, 2019, https://www.ucsusa.org/nuclear-weapons/space-weapons/satellite-database.

195. Ted Muelhaupt and Andrew Abraham, GPS Transponder for Space Traffic Management, August 6, 2018, https://aerospace.org/sites/default/files/2018-08/Muelhaupt_GPS%20Transponders%20for%20STM_0.pdf.

196. David Palmer and Rebecca Holms, Elroi: A License Plate for Your Satellite, February 13, 2018, https://arxiv.org/abs/1802.04820.

197. Joshua Philipp, China is Branding Anti-Satellite Weapons as 'Scavenger Satellites,' May 5, 2019 and updated May 6, 2019, https://www.theepochtimes.com/china-is-branding-anti-satellite-weapons-as-scavenger-satellites_2907825.

scary threat is only a rumor, looking at the progress made by OneWeb and SpaceX in lowering the cost of smallsats and China's significant efforts in smallsats, unmanned robotics, and RPO, it would seem likely that China, as well as Russia, can make numerous cheap robotic ASATs that go to all altitudes including GEO. Such an ASAT does not need to be as cheap as $5 million.[198] Since the typical cost of a US legacy system is on the order of a billion, an ASAT costing $50 million would still be highly favorable to the attacker, even if one ignores the high cost to the victim in losing the services provided by the destroyed satellite until its capability is fully replaced. The United States needs to be prepared to defend especially its expensive and vulnerable legacy satellites that we will still need at least during the 2020s.[199]

D. Other Space Markets

The $1.1-trillion global space industry does not include tourism, resource mining or on-orbit manufacturing. There are indications of the potential values of these space markets. However, since little is known about the costs of capturing those benefits, one cannot determine whether they make business cases or not at this time.

Below, Goldman Sachs provided some highly speculative indications of their potential values.

---

html.

198. BofAML reported that: "Satellites typically cost about US$1bn [billion] to build and launch but the rise of small satellites (CubeSats) means this cost has now been driven down to US$5mn [million] by some companies like Terra Bella/Skybox Imaging. The speed of manufacturing satellites is increasing too; in early 2015 it took Planet Labs just nine days to build two CubeSats. (source: OECD)" (See BofAML, op. cit., p. 62).

199. More about smallsats will be discussed in Section III.C.2. Also, see examples of expensive, vulnerable, but critical satellites during the 2020s in Brian G. Chow, Nuclear vulnerability: In-orbit bodyguards would help protect NC_3 satellites from attacks, *SpaceNews*, April 1, 2019, https://spacenews.com/op-ed-nuclear-vulnerability-in-orbit-bodyguards-would-help-protect-nc3-satellites-from-attacks/.

As to space tourism, Goldman Sachs said that the historical tourist seat on a Russian Soyuz is $35 million per person. Virgin Galactic asks for $250,000 per person for a seat on its sub-orbital spacecraft.[200] However, whether this market would flourish would depend on the safety and fatality records of its flights.[201] It is hard to project its revenue or profit at this time.

As to resource mining, Goldman Sachs cited MIT's Mission 2016 that "successful asteroid mining would likely crater the global price of platinum, with a single 500-meter-wide asteroid containing nearly 175X the global output."[202] Moreover, as water is available from the moon or asteroids, one can use electricity from solar panels to split the water into hydrogen and oxygen and use them as the fuel.[203]

As to on-orbit manufacturing, Goldman Sachs said that using water as an energy carrier for fuel can constitute an orbiting "gas station" and "[a]steroid mining could very quickly supply an emerging on-orbit manufacturing economy with nearly all the raw materials needed."[204] These statements sound even more optimistic than the two above.

---

200. Goldman Sachs, *op. cit.*, p. 3.
201. During an Oct. 31, 2014 test flight, Virgin Galactic's first SpaceShip Two space plane broke up and crashed due to co-pilot error. The co-pilot was killed and the pilot, injured. The Spaceship Two is designed to launch two pilots and six passengers on suborbital rides. (See Tariq Malik, Deadly SpaceShip Two Crash Caused by Co-Pilot Error: NTSB, Space.com, July 28, 2015, https://www.space.com/30073-virgin-galactic-spaceshiptwo-crash-pilot-error.html).
202. Goldman Sachs, *op. cit.*, p. 74.
203. Charles Dunnill and Robert Phillips, Making space rocket fuel from water could drive a power revolution on Earth, the Conversation, September 27, 2016, https://theconversation.com/making-space-rocket-fuel-from-water-could-drive-a-power-revolution-on-earth-65854.
204. Goldman Sachs, *op. cit.*, p. 74.

III. Proposed Measures for Both Economic and Military Security in Space

Brian Weeden reported that "thanks largely to the efforts of Scott Pace and his staff at the National Space Council, who led the interagency efforts that resulted in the first-ever national policy on STM [space traffic management] signed by President Trump last June [2018]." However, Weeden continued that "implementation of that policy has stalled, mainly due to disagreements between Congress [favors Department of Transportation (DOT)] and the White House [favors Department of Commerce (DOC)] over which agency should be in charge."

With the second space age[205] approaching fast, there is no time to waste. This section will describe measures that should be incorporated into STM in order to promote both economic and military security.

A. Establish Red Lines to Assign Liability

National security originally meant military security or protection against military attacks. Over the years, many other dimensions, such as economic, political, ecological, resources, computer and infrastructure securities, have been included in national security.[206] The Directive-3 says "[s]afety, stability, and operational sustainability are foundational to space activities, including commercial, civil, and national security activities."[207] It does not matter whether national security takes the narrowest meaning of military security only or any broader interpretation that includes economic security. That the Directive-3 includes commercial activities in the above state-

205. The second space stage starts when countries begin deploying dual-use robotic spacecraft in the early 2020s and, thus, the space will soon be de facto weaponized. The world must learn how to keep peace and prosperity in a weaponized space.
206. National security, Wikipedia, downloaded on May 1, 2019, https://en.wikipedia.org/wiki/National_security.
207. Space Policy Directive-3, *op. cit.*

ment can be confidently interpreted that safety is foundational to both economic and military security activities. In other words, keeping space assets safe from not just accidental but also intentional collisions is a key goal of the domestic and international STM.

A space object can be a piece of space debris, a functioning satellite or an ASAT. When two space objects interact with each other and one or both space objects get damaged. Let's call this an undesirable space conjunction. How should the liability be assigned? To be more specific, if one of the US satellites is damaged in a space conjunction, the United States needs to react in both a lawyerly way for settling the claim and a military way for protecting the targeted satellite from getting hurt in the first place. The lawyers would want to quickly and clearly assign liability and receive reparation. Understandably, even the quickest way will take months, if not years. On the other hand, in the case that the US satellite or satellites are attacked by ASATs such as robotic spacecraft, the military would need some sort of a red line pre-declared in peacetime. Once that line is crossed, the invading party has committed a wrongful act and the United States can take defensive action to protect its targeted satellite or satellites immediately.[208] That is to say that our military is far more concerned about how to obtain enough warning to activate defense and protect the targeted satellites than getting a monetary settlement. Without these satellites, we might lose the war or, at the least, fighting a bloodier and longer war. This is far more important than getting some money. The warning time to protect targeted satellites is likely to be far more stringent and measured in minutes; unlike a claim settled within months would be considered speedy and satisfactory. In sum, the military wants the targeted satellites to survive far more than it wants a full compensation for the satellites lost.

One method of defense can be the use of bodyguard spacecraft to

208. American Bar Association Standing Committee on Law and National Security and Nonproliferation Policy Education Center, Defending America's Place in Space: Future Threats and Rules, Workshop Report, pp. 11-15, May 2019, https://www.americanbar.org/content/dam/aba/administrative/law_national_security/january-2019-space-report.pdf.

block ASATs from reaching our satellites. Better yet, blocking actions can be performed without hurting the invaders at all.[209]

Directive-3 calls for the establishment of "[a] common process by which individual spacecraft may transit volumes used by existing satellites or constellations."[210] Accordingly, the STM can create "volumes" around those critical but vulnerable satellites that a country wants to protect. The surface of each volume will then be the red line, which instantly defines whether a transiting satellite is outside or inside the protective volume. Then, a common transiting process can be for any adversary to limit the number of transiting satellites inside these volumes not to exceed a certain threshold number.

In November 2018, the *Economist* reported that "Erwin Duhamel, who was until earlier this month head of security strategy at the European Space Agency (ESA), observes that officials in several places are now studying the idea of defending important satellites with 'bodyguard' spacecraft."[211] Some military security planners in the United States and other countries believe that satellite attacks, just like nuclear strikes, can be deterred by using the doctrine of mutual assured destruction. On the contrary, many others including this author argue that effective protection is a prerequisite for a credible deterrence of satellite attacks.[212] It is hoped that satellite bodyguards will soon be recognized as the most effective defense of satellites that can be set up in time against the rapidly approaching robotic threats of the early 2020s.

<u>B. The Directive-3 Tasked DOD to Fill-In the Military Security Blanks</u>

Directive-3 chooses DOC to be in charge of STM, but mandates

---

209. Nuclear Vulnerability, *op. cit.*
210. Space Policy Directive-3, *op. cit.*
211. It will soon be possible to send a satellite to repair another, The Economist, November 24, 2018, https://www.economist.com/science-and-technology/2018/11/24/it-will-soon-be-possible-to-send-a-satellite-to-repair-another.
212. Brian G. Chow, Stalkers in Space: Defeating the Threat, Strategic Studies Quarterly, June 1, 2017, pp. 99-100, https://www.airuniversity.af.mil/Portals/10/SSQ/documents/Volume-11_Issue-2/Chow.pdf.

that "[t]he Secretaries of Defense, Commerce, and Transportation...... shall develop space traffic standards and best practices, including technical guidelines, minimum safety standards, behavioral norms, and orbital conjunction prevention protocols related to pre-launch risk assessment and on-orbit collision avoidance support services."[213] These standards and practices must be specific, transparent and unambiguous so that space users can easily understand and comply with the STM regulations.

For example, as discussed earlier, Directive-3 stated that the United States should establish a process for transiting "volumes used by existing satellites." To make the process enforceable, one must specify the shapes and sizes of the volumes. Say, all those volumes at the GEO altitude are spheres with a radius of 50 km.[214] In fact, the use of specific numerical values in guidelines is common. For example, the IADC Space Debris Mitigation Guidelines use 25 years as post-mission orbital lifetime limitation.[215]

DOD should take charge of determining standards and practices pertaining to military security, because this is the mission of DOD and it has the best knowledge and techniques for their determination including numerically, if it is needed. As it takes time for DOD to choose standards and practices that would produce net benefits to both the United States and its adversaries, DOD should proactively participate in the STM efforts, as opposed to wait passively until it is called upon to do so. Moreover, as military security is not the major concerns of many other agencies, necessary standards and practices, such as the size of the "volume" in Directive-3, pertaining to military security could be ignored or at best inadequately covered if DOD were not taking the initiative to be involved early on. This concern is not hypothetical. While DOD has been focusing on space force

213. Space Policy Directive-3, *op. cit.*
214. Actual number should be determined by DOD after consultation with other agencies domestic and foreign.
215. Inter-Agency Space Debris Coordination Committee, IADC Space Debris Mitigation Guidelines, IADC-02-01 Revision 1, September 2007, p. 9, http://www.unoosa.org/documents/pdf/spacelaw/sd/IADC-2002-01-IADC-Space_Debris-Guidelines-Revision1.pdf.

reorganization issues, there is a presumption that military security can be achieved solely through a space force, while STM needs to focus far more, if not totally, on economic prosperity. Unfortunately, an adversary can hide the preparation of multiple ASAT attacks in the guise of peacetime maneuvers under such a non-military-secure STM. Upon further command at the opening of a space war, the adversary could mount devastating attacks on our critical satellites from such close quarters that we would have no time to defend these satellites.[216]

Another organization that can be easily overlooked by DOD is the Consortium For Execution of Rendezvous and Servicing Operations (CONFERS), an industry-led initiative with initial seed funding provided by DARPA. It "aims to leverage best practices from government and industry to research, develop, and publish non-binding, consensus-derived technical and operations standards for OOS [On-Orbit Satellite Servicing] and RPO [Rendezvous and Proximity Operations].[217] It issued its recommended design and operational practices on February 1, 2019. It echoed Directive-3 by saying that "specific techniques [for spaceflight safety] may include passive safe orbits, safety zones, and keep-out spheres or volumes for RPO and OOS activities."[218] Again, unless DOD takes an active role, this industry-led initiative is likely to favor economic prosperity and may neglect military security. Moreover, its "non-binding" and "consensus-derived" standards may not be followed by an adversary at time of crisis or war when its national interest differs from ours.

---

216. Brian Chow, The Greatest Threat to America's Military? A 'Pearl Harbor' In Space, the National Interest, July 6, 2018, https://nationalinterest.org/blog/buzz/greatest-threat-americas-military-pearl-harbor-space-25142.
217. The Consortium for Execution of Rendezvous and Servicing Operations (CONFERS), CONFERS, downloaded on May 7, 2019, https://www.satellite-confers.org/about-us/.
218. CONFERS Recommended Design and Operational Practices, CONFERS, February 1, 2019, https://www.satelliteconfers.org/wp-content/uploads/2019/02/CONFERS-Operating-Practices-Approved-1-Feb-2019-003.pdf.

C. Liability for Damage Caused by a Space Object

If our satellite is damaged by another satellite, whether accidentally or internationally and whether owned by a domestic or foreign company, the goals should be:

- The at-fault party or the wrongful act is quickly and clearest determined by regulations or laws.
- The United States can protect the targeted satellites from damage.

This article suggests measures below to meet each goal.

1. Determine At-Fault Party or Wrongful Act Quickly and Clearly

A domestic or international STM would have to deal with liability for damage caused by a space object. Article VII of the Outer Space Treaty of 1967 states:

> "Each State Party to the Treaty that launches or procures the launching of an object into outer space,…and each State Party from whose territory or facility an object is launched, is internationally liable for damage to another State Party to the Treaty or to its natural or juridical persons by such object or its component parts on the Earth, in air space or in outer space."[219]

The treaty assigns the liability for damage by the space object to the country, as opposed to the owner. This article recommends, that as the space industry matures and grows into a trillion-dollar market by 2040, the liability, just like those from other mature industries, should eventually be shifted to owners of the space object or its component parts that causes the damage.

219. United Nations Treaties and Principles on Outer Space, United Nations 2002, p. 6, http://www.unoosa.org/pdf/publications/STSPACE11E.pdf.

However, since the space threat environment requires a functional STM by the early 2020s, the switch of liability from country to owner cannot be done in time. The United States and the world would have to rely on the Liability Convention for the 2020s and, thus, buy time for the eventual switch.

Article III of the Liability Convention states:

> "In the event of damage being caused elsewhere than on the surface of the Earth to a space object of one launching State or to persons or property on board such a space object by a space object of another launching State, the latter shall be liable only if the damage is due to its fault or the fault of persons for whom it is responsible."[220]

Let's apply Article III to a key concern of this article, namely, the liability for damage, whether accidental or intentional, to a US satellite already in orbit by a Chinese robotic spacecraft. Under Article VII of the Outer Space Treaty, China is liable. However, Article III of the Liability Conventions states that China is only liable if it is at fault. Moreover, the time to determine whether China is at fault and how much the United States should be compensated may take three-and-half years.[221] That is a long time for the United States to sue, prove China to be at fault, and finally get compensated.

There is even a far more fundamental problem with Article III. Article II of the same Liability Convention says:

220. Ibid., p.15.

221. Article X of the Liability Convention says that a claim should be filed not later than one year following the date of the occurrence of the damage. Article XIV says that, "[i]f no settlement of a claim is arrived at....within one year from the date on which the claimant State notifies the launching State...., the parties concerned shall establish a Claims Commission." Article XV adds six months to appoint the Chairman of the Commission, if the claimant State and the launching State cannot agree on the choice of the Chairman. Finally, Article XIX states that the Commission shall give its decision or award no later than one year from the date of its establishment. Thus, the total amount of time is up to 3.5 years. (See Ibid., pp. 17-20)

> "A launching State shall be absolutely liable to pay compensation for damage caused by its space object on the surface of the Earth or to aircraft in flight."[222]

The difference between fault-based liability in Article III and absolute liability in Article II is damage due to mistake (i.e. inadvertent action). While damage due to fault (i.e. intended wrong action) would have to be compensated under fault-based liability, mistaken action escapes liability. On the other hand, absolute liability calls for compensation regardless mistake or fault. Perhaps, when the Liability Convention was considered and negotiated during 1963-1972, people wanted absolute liability for damage to humans "on the surface of the Earth or to aircraft in flight" as called for in Article II. Since there were few humans (astronauts being the exception) in space, there was little need to include absolute liability for damage in outer space in Article II. As humans travelling in space will be far more common, absolute liability should be applied to Article III as well so as to compensate human loses in space, just as "on the surface of the Earth or to aircraft in flight."

The United States should propose an amendment to merge Article III into Article II by making liability absolute regardless damage caused by space object on the Earth or in outer space. In other words, Article III will no longer be needed so long as Article II reads:

> "A launching State shall be absolutely liable to pay compensation for damage caused by its space object on the surface of the Earth, to aircraft in flight or in outer space."

With the above amendment, the liability will be the same regardless where the damage occurs. Such an amendment can not only fix the fundamental problem that the liability should be the same regardless of whether the damage occurs on Earth or in outer space, but also eliminate the time needed to determine who is at fault and allows the settlement of the case quicker and with less effort.

On the other hand, it still takes time to determine whether the dam-

222. Ibid., p.15.

age to our satellite is caused by a wrongful act, which is still required for absolute liability. As will be discussed below, this determination can be made quickly.

Scott Kerr, as well as many other scholars, argued that "it is quite likely that many space-related international treaties (such as the Outer Space Treaty and the Space Liability Convention) have crystallized into customary international law."[223] Accordingly, the United States should incorporate, as a red line, the aforementioned (in Section III.A) concept of "volumes used by existing satellites" indicated in Directive-3 or "safety zones, and keep-out spheres or volumes" described in CONFERS into its space policy; and take the lead to include the same concept, albeit necessarily in various levels of specificity, into international transparency and confidence-building measures, guidelines, and STM. The United States should also practice, regulate and enforce the rules based on the red line and encourage other countries to do the same. These activities will help turn these rules into customary international laws in space.

Leon Castellanos-Jankiewicz stated that "damage and fault are not required elements of international State responsibility, pursuant to the objective nature of the internationally wrongful act, which only necessitates a finding of wrongful conduct attributable to a State in order to arise."[224] Since crossing a red line is observable, the first goal of quick and clear determination of wrongful act is attained.

2. Protect Our Satellites from Robotic ASATs

Unfortunately, even an instant determination of wrongful act is not sufficient to protect our satellites. In a contingency of China taking Taiwan by force, China still has large incentives to threaten our sat-

223. Scott Kerr, Liability for space debris collisions and the Kessler Syndrome (part 1), December 11, 2017, http://www.thespacereview.com/article/3387/1.
224. Leon Castellanos-Jankiewicz, Causation and International State Responsibility, Amsterdam Center for International Law, April 3, 2012, p. 4, file:///F:/NPEC%20space%20project/reports%20to%20copy/Castellanos-Causation-and-International-State-Responsibility1.pdf.

ellites when its national interests loom large. Starting its preparation months before a planned contingency, China could have pre-positioned an arbitrarily large fraction of its robotic servicing spacecraft fleet, as well as in the future robotic smallsats,[225] arbitrarily close to our critical satellites at the eve of a major crisis. China is counting on one of highly favorable outcomes.

First, if the United States intervenes, these robotic spacecraft already positioned closely would move the last short distance and disable those targeted satellites so quickly that the United States could not activate its defenses in time to protect these targeted satellites. Without the support of these satellites, our warfighting would be ineffective and the intervention, fail. Then, even if China had to compensate the United States for destroying our satellites, the payment would pale in comparison with the prize of China's successful unification. Better yet for China, the United States, after losing a war, could have so many bigger issues to deal with rather than wasting the time and energy to negotiate with the victorious China for a relatively little reparation for the losses of satellites.

Second, knowing its intervention without the support of these critical satellites could fail or, at best, be far bloodier and longer, the United States might decide not to intervene at all. Then, China would deter the intervention without firing a shot and, to boot, does not incur any liability, as none of our satellites is damaged.

Bodyguard spacecraft can protect our satellites against these robotic threats. The use of bodyguards in space is a logical extension of our common practices in the air, at sea and on land. Countries including the United States use guards to protect their vital and vulnerable assets. Yet, for the last six decades, the world has been accustomed to and spoiled by a peaceful outer space, and believes that guards and space weapons are unnecessary for protection. The rapidly approaching second space age will be unavoidably and irreversibly weaponized by the presence of dual-use robotic spacecraft. On the other hand, many are still living in the dream of a weapons-free

---

225. See more detailed discussion of these rumor Chinese smallsats for ASATs in Section II.C.

space. A practical compromise for them is to negotiate for a minimization of space weapons but, at the same time, realize that it is impractical to avoid the use of some defensive weapons to counter ASATs.[226]

Section III.A has already noted that bodyguard spacecraft do not need to harm invaders inside the red line. Bodyguards can maneuver to block the invaders from reaching the critical satellite(s) in the zone. This defense would be the least escalatory and most proportional response to these threats.

Section II.C pointed out that, sometime in the 2020s, China and Russia will likely be capable to produce a large number of inexpensive robotic ASAT smallsats. The concept and rules of a red line will continue to be necessary in countering this far more serious robotic threat. Also, the United States can similarly use cheap smallsats as bodyguards to protect our satellites.

D. Pushing for the Same Indemnification Provisions Among Countries

If a domestic or foreign spacecraft, during its transit or service, accidentally or intentionally causes damage to a US satellite, it is only fair that the responsible party (i.e. first-party) of the liable spacecraft, as opposed to that (i.e. third party) of the US satellite, should have carried an insurance to pay for the damage. This is the purpose of third-party liability insurance.[227]

In November 2015, the US Congress approved the Spurring Private Aerospace Competitiveness and Entrepreneurship (the SPACE Act

226. For a comprehensive list of space arms control measures to more effectively manage both traditional and emerging space weapons, see Brian G. Chow, Space Arms Control: A Hybrid Approach, *Strategic Studies Quarterly*, May 31, 2018, https://www.airuniversity.af.mil/Portals/10/SSQ/documents/Volume-12_Issue-2/Chow.pdf.

227. In third-party liability insurance, the insured is the first party. The insurer is the second party. The party that files a claim against the insured is the third party.

of 2015). The Act states that "[t]he liability coverage of licensees subject to third-party claims exceeding the amount of insurance or demonstration of financial responsibility shall be extended through FY2025."[228] The United States will indemnify US space licensees' third-party losses for bodily injury or property damage above the "maximum probable loss" (MPL) from a permitted or licensed activity. Milton "Skip" Smith provided details to the Act. He stated:

> "The licensee or permittee must obtain third party liability insurance or demonstrate financial responsibility in amounts sufficient to compensate for the MPL.[229] The third party MPL amounts are established for each license by the FAA [Federal Aviation Administration] up to a maximum of $500 million or '[t]he maximum liability insurance available on the world market at a reasonable cost.'[230] Similar provisions apply to claims by the United States, its agencies, and its contractors and subcontractors, with their MPL capped at $100 million or '[t]he maximum liability insurance available on the world market at a reasonable cost.'"[231]

Under this Act, licensees would have incentives to strive for a better safety record in order to lower the premium for the third-party liability insurance. Further, licensees would want to observe the rules according to a red line, because by doing so would help avoid coming too close to, and running into, third-party satellites.

However, countries have been establishing their own indemnification and other provisions for third-party liability insurance. If a Chinese robotic RPO spacecraft damages, whether accidentally or intentionally, a US satellite, it would be unfair that the United States would be compensated according to the provisions of China. The United States should push for the same provisions for all spacefar-

---

228. H.R. 2262 - US Commercial Space Launch Competitiveness Act, downloaded on May 15, 2019 from https://www.congress.gov/bill/114th-congress/house-bill/2262.
229. 14 Code of Federal Regulations Section 440.9(a) and (b).
230. 14 Code of Federal Regulations Section 440.9(c)(1) and (2).
231. 14 Code of Federal Regulations Section 440.9(d) and (e).

ing nations in international fora.

V. Conclusion

This article has found that, in the next two decades, the United States will still have the largest market share in practically every space industrial sector, such as launch, satellite manufacturing, and, the largest sector of all, satellite services. On top of that market power and influence of the United States, European countries and other western countries will continue to capture sizable market share in all space sectors, leaving China and Russia very weak in commercial space. The US should proactively develop, practice, and demonstrate norms, standards, regulations and laws in its space commerce and the domestic STM. The United States should also take the initiative to work with other western countries and whoever else wants to join and use their domestic STMs as a basis to establish a multinational STM. The United States should use the same principles in the multinational STM to simultaneously lead the way for an international STM. Both Russia and China are not equipped to compete with the West in a free and competitive market, especially the rapidly expanding satellite services sector that favors quick and frequent innovations, large capital investment, and those capturing the market first. Thus, forming partnerships with the West is likely to be the most viable route for them to participate more meaningfully in the economic prosperity bestowed by the growing global space industry. But then, they would have little choice but fall in line with the STM designed by the West.

This article has proposed five measures as building blocks for developing standards, practices, regulations and laws in STM so as to meet the dual goals of economic prosperity and military security:

1. Establish red lines in space surrounding critical satellites to quickly and clearly assign liability, when an undesirable space conjunction occurs.

2. With the best data and skills, DOD should take the initiative to develop specific space traffic standards and best practices per-

taining to military security in STM. Otherwise, economic agencies and commercial operators could favor economic prosperity over military security.

3. The United States should submit an amendment to the Liability Convention to change the current fault-based liability for damage in space by a space object to absolute liability, just the same as liability for damage on Earth by a space object. The change would also facilitate the rules pertaining to the red lines in 1. to become customary international laws, regulations, and enforcement in STM.

4. The United States should deploy bodyguard spacecraft to get ready in time to protect satellites against the rapidly emerging and growing robotic ASATs.

5. For fairness, the United States should take the lead to make all spacefaring nations have the same indemnification and other provisions in their third-party liability insurance.

The second space age will be here by the early 2020. this does not leave much time to have a functional STM to assure a continuing peaceful and prosperous space.

Chapter 4:

Space security in the 21st century: a French view

Xavier Pasco

I - Introduction

At the occasion of the Bastille Day 2018, and in the aftermath of "National Security and Defense review", President Macron ordered the Minister for the Armed Forces, Florence Parly to produce a new "Space Defense Strategy"[232] with the objective to prepare France for a continued independent use of space for its own national security and to address the new challenges raised by this transformed space environment. This document prepared over an 10 month period was finally published in the summer of 2019 and comprises a number of new announcements for future military space never specifically addressed ever before.

More particularly, an important part of the new strategy is devoted to the new space context (including the development of space threats and risks) and, for the first time, provides details about future military uses of space that go beyond the classical Earth observation, telecommunication or timing and positioning. In an accompanying speech, Florence Parly was not shy to address the new concept of "active defense" in space, stressing in particular the need for France to get equipped with specific space weapons, even if this particular expression has never been used. This new situation has put France under the headlights, notably in a context where discussions about the development of safer, more secure, and more sustainable devel-

232. Space Defense Strategy, Report of the " Space " working group, 2019, DICOD, French Ministry for the Armed Forces, November 2019, p. 59.

opment of space activities have been marked by string divergences between the main space powers in the United Nations or in other forums.

This piece is intended to analyze the new French posture, provide elements about its organizational and programmatic consequences and provide a policy-oriented analysis of its objectives. Before addressing the current thinking, the paper takes care to put this new discourse in perspective and provide some historical hints that has made France particularly active in military space for decades at the difference of its European neighbors for example. It is suggested that this specific positioning largely explains the French perceptions of the current space dynamics and the new posture it has decided to publicize. None the least of the challenges will consist in combining this new posture with a dynamic diplomatic engagement pleading for creative mechanisms that promote a larger and more effective collective security in space. This paper will attempt to conclude on possible solutions in this regards and will argue about possible positive outputs of a carefully weighed national doctrine that may hopefully prove a pragmatic possible step in this direction.

II – The Roots of French Military Space: Recalling the Nuclear Deterrence Context

The history and development of military space in France has reflected a unique and singular posture in Europe after the war. The early links with the construction of the national nuclear deterrence force have formed the backbone of the national military space effort. Over the years, the political need for affirming a high level of national independence in this strategic domain has only increased and has rendered the mastering of strategic space systems necessary. This has been especially the case when the status of the nuclear deterrence was challenged in the eighties by the advent of a new strategic posture in the US with the advent of the "Strategic Defense Initiative" (SDI) initiated by Ronald Reagan. Dubbed "Starwars," the proposed concept intended to deploy a strategic shield against incoming Soviet Intercontinental missiles, thus potentially rendering nuclear mutual deterrence obsolete as a strategic concept. France, a nuclear power itself, was most attentive to the coming SDI developments and space systems clearly emerged as a key element of the project.

More generally, this strategic-led environment has been both structuring and constraining regarding the developments of military space systems during this period. The green light for the development of military observations or telecommunication satellites was given with the perception that it would increase the efficiency of the nuclear deterrence. This has been also the time when the HELIOS program, a satellite series for military observation and targeting, has been decided (1986). Still, this development was merely conceived as a peripheral element (as well as space telecommunication satellites) in the deterrence context without giving birth to an all-out development of other space systems. For example, no space-based early warning systems was planned at that time. The main preoccupation was to highlight the value of the French nuclear retaliation doctrine as it existed without giving the feeling of advocating for new defensive developments at the very moment when the US were choosing another path by advocating the end of the MAD era with the SDI and its evolutions in space.

However, there is no doubt that space gained clearly a new strategic visibility for French political and military authorities during the eighties, whether for reinforcing the deterrence policy or for assessing the level of possible new defensive systems against long-range ballistic missiles. Besides recalling the close link that has been established very early in France between the nuclear deterrence and space, looking back at this period also shows why space systems have immediately gained a very highly valued status in the view of the successive French policy makers. It has progressively become clear that such rare systems should be considered as critical and should be supported and protected. And even if security in space was not viewed at the time as a first-rank challenging issue (very few national actors were then able to put satellites in space), French military satellites have progressively appeared as crucial in the French general strategic and military posture.

III – Transforming the Reliance on Space : "Knowledge and Anticipation"

Many transformations have affected affect the national military space posture and have led to the consolidation of the role of military space systems in the 2000s. In particular, new forms of threats were emerging and were calling for a deep reshaping of the military posture. Two main areas of improvements were then identified: Intelligence collection capabilities and the preparation of the military forces for possible long-range external operations destined to thwart mounting terrorism. After a few years of debates, the "Defense and Security White Book" published in 2008 under the Sarkozy presidency has been the first official document to take this new dimension into account by creating a so-called "fifth strategic function" titled "Knowledge and anticipation." The objective was to better prepare France to new forms of security and military threats. Among other measures, it promoted an increased use of space systems for intelligence and for supporting combat operations. This enlarged use of space systems beyond the only strategic intelligence level meant a whole new role for satellites with a more operational theater-level function. It would imply the development of new capabilities (including for example the beginnings of early warning experiments or electronic intelligence orbital systems) that would better serve military operations. It has also led to the decision to create a new "Joint Space Command" whose main function would be to coordinate the use of military systems by the armed forces. This new Joint Command made its debut in 2010.

A renewed version of the "Defense and Security White Book" was published in 2013 reaffirming the role of space in the military posture. In particular, new possible developments were this time mentioned regarding the use of early warning systems which would show how much the doctrinal thinking had evolved since the first decisions made in the eighties (such functions had been kept out of the realm of possible space missions at the time).

Last but not least, the 2017 "Strategic Review of Defense and National Security"[233] ordered by President Macron has definitely put space at the forefront of the national military posture. It has first acknowledged the crucial role played by space systems in most of the recent external military operations. The example of "Operation Serval" in Mali had then largely been seen as confirming how much using modern and adapted military and dual-use space systems could improve military efficiency when confronted with large area surveillance and combat operations.

But this document also indicated for the first time that, given this new dependency, France should prepare to protect its military satellites against the emergence of new threats in space. This was really new. It has resulted from a growing perception that space was getting transformed and was gradually appearing as a possibly conflicting environment. The document made it explicit: "The problem of space weaponization must be now addressed in new terms."[234] Since the Chinese ASAT test in 2007 followed by early reaction by the US and by many Russian and Chinese new military experiments, the feeling has increased that military dynamics in space would get more uncertain and more unpredictable. To complete the picture, the emergence of many new actors, including private ones, with many new small satellite projects, had added to the uncertainty for the military users of space. France would then have to prepare to face new risks and threats.

IV – Towards a New French Defense Posture in Space

A new "Space Defense Strategy" was published in the summer of 2019. It first announced the creation of a new French "Space Command" giving the Air Force a clear command and operational lead in the field (the French Air Force is to be renamed "Air and Space Force"). This creation is expected to mark both a qualitative and a quantitative improvement over the existing Joint Space Command.

---

233. The document can be found at https://cd-geneve.delegfrance.org/Strategic-review-of-Defense-and-national-security-2017.
234. Idem, p.45.

It first intends to send a strong message regarding the new policy approach adopted by France about the military use of space. As any reader can remark, a first part of the Space Defense Strategy as well as major moments in the speech made by the ministry presenting this new strategy on July 25, 2018, in Lyon[235] have been devoted to highlighting new threats France would have to face in space in the coming years. The document announces it clearly in the opening pages and then makes it a further key subtitle: "Strategic competition is heating up, operational contexts are evolving,"[236] then, "French actors in both the public and the private sector must guard against more recent threats, such as the proliferation of space debris, jamming, blinding and directed-energy weapons."[237] In her speech, Florence Parly herself chose to insist on what she calls a "ramping up weaponization" in space: "Today, our allies and our adversaries have been militarizing space. And as the time for resilience is becoming ever shorter, we have to act. We must be ready. Indeed, tomorrow is already yesterday. (…) Satellites spied on, jammed or blinded; means to disturb, neutralize or destroy space capabilities exist and they develop: we know it, the shadow of the threat is real for sure."[238]

Care has clearly been taken to link this approach to the increasing dependency on space for the military users. Giving a new role for the military in space derives firstly from the reinforced status given to space in the larger French defense posture for carrying out its military operations on the ground: "Our first responsibility is to protect our space capabilities. They are essential to our operations, they are essential to the functioning of our economy and of our society."[239]

In this context, the French Space Defense Strategy can be viewed as being structured around two main interrelated messages. First, it

235. See https://www.defense.gouv.fr/content/download/563595/9727199/Discours de Florence Parly%2C présentation de la stratégie spatiale de défense à Lyon%2C le 25 juillet 2019.pdf.
236. Space Defense Strategy, p.9 and 21.
237. Idem, p.9.
238. Ministry speech mentioned above, p.2 and 3.
239. Idem, p.8.

clearly states the need for reinforcing the military capability to operate in space if needed by giving a more operational role to the military, including in operating their own satellites (this role was previously delegated to the French space Agency, CNES). But it also strongly underlines the reinforced link between this space reorganization and the larger operational role given to space capabilities over the recent years. This is the reason why the new Space Command has clearly been advocated by the Minister herself as placed under the responsibility of the Air Force: "Space is a natural extension of the air domain, a field in which you [Air Force] are experienced. This is why this great command will be place under the authority of the Air Force, which will thus become the Air and Space Force."[240] Do not mistake. This message that many US airmen may have liked to hear from their own political authorities while they were thinking about creating separate new "Space Forces" is not meant to merely satisfy corporate ambitions. From a political standpoint, it is mainly destined to reaffirm space as an operational resource before being a warfighting domain. In other terms, it may become the latter if and only if it cannot render the operational service it is usually used for. In this logic, there was certainly no need for developing a separate "Space Force" which did not have any advocate in France, by contrast with the US for example. This quest for a strong military policy coherence has seemed to be widely accepted as a prerequisite for any development of a new space-oriented military organization. It has also allowed introducing new functions for this organization, including the mission to operate its own capabilities but also the possible fielding of new "active defense" space systems.

Consequently, once accepted, the protection of one's own capabilities and the probability to implement in space any required military maneuver gained legitimacy and would thus imply a change in chain of command for satellite operations. This mainly explains the significant change of size that has been planned in term of staffing. From a few dozens of military personnel, the space command shall count a few hundred in the months to come, possibly increasing in a

240. Idem p.4.

next step up to about 500 personnel at cruise speed.[241]

V- "Active Defense" and Lasers in Space

But of course, the main new function would deal with implementing an "active defense" in space which Minister Parly has clearly evoked in her speech:

> "This is why I have decide to launch a new armament program called 'Mastering space.' Just that. It will comprise two parts: surveillance and active defense. Our first responsibility is to protect our capabilities in space. (…) We have to watch our satellites more and better. We have to know perfectly the objects that surround them, which cross their trajectories. We must be able to detect and attributes the suspicious, unfriendly, or even hostile acts undertaken against our military satellites and our interests in space. (…) Once equipped with these new surveillance capabilities, we will be able to organize our active defense. And here I want to be precise: Active defense has nothing to do with an offensive strategy, it is rather about self-defense. It is about, when a hostile act has been detected, characterized and attributed, responding to it in an adapted and proportionate way, in compliance with the international law."

Following these comments, Minister Parly particularly referred to ground-based or even space-based lasers embarked upon "precious" satellites and to "patrollers nano-satellites" envisaged to better monitor possible events or attacks and undoubtedly pave the way for further response.

The possible use of laser technology, whether ground-based or embarked, has obviously not been mentioned by chance. Lasers belong to the so-called directed energy technologies that present two main

241. As mentioned in an interview given by F. Parly. on 24 July 2020 in the French press.

characteristics when associated with their use against supposed space targets[242].

First, they do not imply hard hitting the target and their use does not create multiple debris. By contrast, kinetic energy weapons in space, such as ground-based missiles that have been used by several space powers in history and more recently, do create debris in space. While some of those countries have argued that such tests have been performed in a controlled way to minimize the pollution in space, it remains that such a performance may be discussed and bring about a dangerous trend by itself. For a few years already, the space environment has been radically transformed with an increased number of new actors sending and operating a huge number of space objects, a situation never experienced before. The years to come will just vastly increase the number of satellites by order of magnitudes. From a mere 1000 functional satellites that could be counted in space a few years ago, the number will soar to dozens of thousands of satellites, leading in a few years to a situation where there will be more active objects in space at a given time than the total number of satellites that have been launched from 1957 on. In such a situation, any direct contribution to increasing the level of lethal debris in space does seem politically irresponsible. The probability of collisions with debris will already naturally increase and it is worth recalling that even small-size debris, i.e. non detectable few centimeters debris, can be lethal for any satellite. Besides killing satellites, such collisions will also bring their tribute to the production of new debris, starting the well-know cascading effect feared by all the orbital mechanics specialists and known as the Kessler syndrome effect. Thus from a mere practical standpoint, and besides the bad political message regarding the perspective of a "safe, secure and sustainable space environment," the use of kinetic energy weapons producing long-lived debris clearly appears counterproductive, to say the least, in the perspective of a sustained military use of space.

---

242. For a more complete perspective on the technology families possibly used as space weapons see Pasco (Xavier), *Various Threats of Space Systems*, in Schrögl (Kai-Uwe), Ed., *Handbook of Space Security: Policies, Applications and Programs*, Springer Verlag, 2020.

Space physics does not reserve special treatment to military objects and regarding such hits as possible standard practices for the military would eventually constitute nonsense.

Obviously, some other "kinetic" techniques such as "rendez-vous" and controlled direct contact with space objects may certainly be envisioned by technicians. After all, several programs exist that use maneuverable spacecrafts that can inspect or even intervene on satellites. For example, the US MEV-1 technology experiment implying the docking of a spacecraft designed by Northrop Grumman to a commercial telecommunication satellite (Intelsat 901) has been successfully carried out last February. The objective in this case was to allow an extension of life of Intelsat 901 for another 5 years by MEV-1. NASA has also supported such research with its Restore-L program aiming this time at demonstrating the feasibility of refueling and assembling missions in Low-Earth orbits (LEO). In Europe as well, demonstrators of maneuverable "spacetugs" have been envisioned, both for allowing satellite life extension or repairs but also in the context of debris remediation. This activity has notably been supported by the European Space Agency (ESA program e.Deorbit). The European experience with the ATV (Automated Transfer Vehicle) used for the service of the International Space Station is of course one of the building blocks of such activities. The need for "active debris removal" technologies (often known as ADR) makes such missions legitimate for ESA to undertake.

But other types of unauthorized rendez-vous and proximity operations may exist and, by the way, during her July 2019 speech, Florence Parly has not been shy talking about the suspect behavior of the "indiscreet" Russian satellite Luch-Olymp, that "had left its visit card to 8 satellites" according to the French minister.[243] A number of other space rendez-vous, proximity operations and even contact experiments, notably attempted by China for more than five years now[244] have taken place for the last few years.

243. Idem, p.3.

244. All these experiments and maneuvers have been regularly well documented in the annual "Global Counterspace Capabilities, An Open Source Assessment"

More generally, all these operations have a dual-capacity and may serve civilian missions as well as military operations. It is also true that the advancement of the mega-constellations projects possibly leading to several thousand objects in LEO will make necessary new techniques for safe disposal and repair of those many satellites. But this increased space traffic will lead to more uncertainties, possibly more misunderstandings, especially in a context where military space operators would attempt to multiply such experiments... By explicitly referring to lasers, it seems like the French MoD wants to underline the "responsible" stance it intends to promote for its new space doctrine. By investing on such techniques, the new French defense space policy wants to be coherent with the traditional support brought by France to the United Nations effort of a "safer, more secure and more sustainable space"... while attempting to demonstrate that is does not necessarily mean renouncing to national defense in space. Quite a challenge to meet.

VI -A Renewed Political Vision of the "Space Dilemma"

For sure, none of these techniques, whether directed-energy or kinetic energy-based, contravene to the letter of the 1967 Outer Space Treaty (OST), the principal legal instrument for space activities signed by the main space faring countries and that remains today the key document of the international space law. However, while the compliance with law is a key element, this does not mean that the spirit of any decision cannot be contested with regard to its common understanding by the international community. After all space has been seen as immune from weaponization for decades and such decisions can appear as giving a first blow in the collective understanding of what is a decent behavior in space. Reconciling the need for legitimate space defense (as pursued by the French space doc-

---

report edited by the Secure World Foundation. See the latest version dated April 2020: see https://swfound.org/counterspace/. Incidentally, it must also be recalled that China largely contributed to the resumption of hard-hitting kinetic tests after having destroyed one of its old satellites with a missile in January 2007, leading to several thousands of debris, most of them still in orbit.

trine) and the spirit of the peaceful uses of outer space as proclaimed in the OST constitutes what may be called a "space dilemma". For sure, insisting on the use of directed energy weapons, as promoted by Florence Parly, may be seen as a first, possibly partial answer to this dilemma.

This view can be supported by a second characteristic inherent to the use of such technologies. Again by contrast with most of kinetic energy weapons demonstrated so far (i.e. co-orbital anti-satellite tests or missile interceptions), laser or other directed-energy technologies may not necessarily imply an irreversible destruction of the target in space. In particular, given the type of laser source utilized, the level of power involved and the duration of the operation, these new techniques may provide some operational flexibility and provide a form of control on the final effect. This is of course of particular importance for the defender to select the kind of action it wants to undertake, keeping it for example proportionate to the seriousness of the hostile action. This is the "adapted and proportionate way", mentioned by the French minister and quoted above, "in compliance with the international law," as she pursued. Indeed, depending on how they are used, lasers can only temporally disturb space systems, blind observation satellites for example or induce other effects that can be reversible. It is easy to understand that such an operational flexibility may provide a highly desirable flexible solution both from a military and from a political standpoint.

All this is not new for sure. Military interest for such adaptable techniques has been expressed already in early 2000s documents published in the United States. An Air Force doctrinal document published in 2004[245] was introducing notions like "defensive counterspace" (interestingly defined as the "ability to exploit space to its advantage via active and passive actions to protect friendly space-related capabilities from enemy attack or interference") or "offensive counterspace" (aiming at precluding "an adversary from exploiting space to their advantage"), already paving the way for bolder mili-

245. See Counterspace Operations, Air Force Doctrine Document 2-2.1, 2 August 2004.

tary activities in space. From then on, a number of analyses have pointed out possibility, for enemies as well as for allied or friendly forces, to use discriminate levels of action in space: "Deception, disruption, denial, degradation and destruction, commonly known as the 'Five Ds' are the possible desired effects when targeting and adversary's space capability" as it was recalled in the offensive space control section of the annex 3-14 of the 2012 Air Force operational level doctrine.[246] The 2018 updated version goes further by mentioning the reversible (temporary) character of the three first forms of actions ("Deceive, Disrupt, Deny") and the irreversible (permanent) type of the two latest ("Degrade, Destroy")[247].

In a sense, this "flexible response" capability (to use a well-known expression in a different context) clearly offers the room for a genuine political handling of possible space confrontations. In other terms, by allowing tailored military decisions, the perspective to use such technologies has introduced a new political dimension to possible space operations. The French doctrine has clearly taken stock of those accessible technologies and it is not by chance that for the first time a genuine high-level political posture proposing a "defense space strategy" has been articulated in France. Indeed, it may well be because such technologies may offer this range of actions that military space has probably entered in a new political era that this new French policy has been, in a way, giving credit to.

VII -The Difficult Art of Making Space Defense Real

Also this new political dimension of military defense as related to the possible use of new "counterspace" techniques has been reinforced by the possibility given to exert defense in space in a politically acceptable way. More traditional and destructive anti-satellite techniques cannot enjoy the same level of acceptability, legitimately

---

246. The 2012 version of the AFdoctrine can be retrieved at: https://fas.org/irp/doddir/usaf/3-14-annex-osc.pdf.
247. See https://www.doctrine.af.mil/Portals/61/documents/Annex_3-14/3-14-D05-SPACE-Counterspace-Ops.pdf.

so. On the latter, the French strategy makes it clear that hard-kill intercepts by ground-based missiles as they have been attempted so far cannot be supported any longer: "France could support the promotion of a norm prohibiting actions that create pieces of long-live debris."[248]

One can only remark how much such a strong statement on kinetic-energy systems contrasts with the push for the use of lasers in the minister's speech :

> "If our satellites are threatened, we will consider dazzling those of our adversaries. We reserve the timing and means of response: this may involve the use of high-powered lasers deployed from our satellites or from our patrol nano-satellites. Because, yes, we will develop high-power lasers. This is an area in which France is lagging behind. But we will make up for it, and I hope that we will be able to equip our precious satellites or our nano-satellites with them as soon as possible to keep at a distance and, if necessary, dazzle those who might be tempted to get too close. So, as you will have seen, the program is vast and ambitious. It will concentrate all our energies on ensuring that we have the initial capabilities required under the current military programming law for full capability on the horizon 2030."[249]

This strong posture easily shows a willingness to send a clear signal that defending satellites can be politically endorsed and thus actually exerted. This signal is obviously key as it may also be seen as evoking in France the beginnings of what has been sometimes referred to as a "space deterrence " approach often debated in experts' circles, especially in the United States.[250] For sure France does not

248. Space Defense Strategy, p.27.
249. Ministers' speech, 28 July 2018.
250. See for example first analysis such as Harrison R., Jackson D., Shackelford C., *Space Deterrence, The Delicate Balance of Risk*, Eisenhower Center for Space and Defense Studies, 2010, 30p; Marquez P., *Space Deterrence, The Prêt-à-Porter Suit for the naked Emperor*, in *Returning to Fundamentals: Deterrence and U.S. National Security in the 21st Century*, The Marshall Institute, 2011,

refer to the deterrence concept in her current space defense posture. In the first place, the very notion of deterrence remains in France almost exclusively linked to the nuclear posture. Only if for preserving this key element of the French national defense policy from further evolving internal debates, extending the notion to other domains seems to remain very unlikely for the years to come.

But beyond this particular "cultural" aspect, using the ambitious deterrence concept in space may remain hazardous to some degree. First, the context of effective deterrence implies mutually understood definitions and clear adherence to norms of behaviors. Common perceptions of such notions as "vital interest" or "red lines"; of procedural steps leading to chains of consequences are key to the effectiveness of deterrence. The very concept of "nuclear deterrence" has evolved over the cold war years almost in a theoretical way in relation with heavily codified sets of rules in the context of a highly specific strategic environment. It remains to be seen how much such a concept may apply to a much different space situation. Space hardly appears as a homogeneous place where capabilities and actors are equivalent in volume and nature. Only the difference in national policies in terms of reliance of space necessarily induces different perceptions on the value of space assets ad it is not by chance that the US has been the first to tackle the issue of "space deterrence." Last but not least, the relatively incomplete character of the global "space picture" (or Space Situational Awareness –SSA- as known in the military jargon) does not seem to provide the level of certainty required for effective deterrence. The attribution of a negative effect to a positively asserted cause remains the space decision-makers' *Holy Grail*. In a space medium that has often been described as "contested, congested and competitive," discerning with precision the origins of an orbital accident and the actual intent of a detected action will in fact become a more and more challenging task.

Also, the notion of "space deterrence" may get complicated by its dilution in larger space security related discussions that it is part

---

pp.10-19; Morgan F., *Deterrence and First-Strike Stability in Space, A Preliminary Assessment*, Rand Corporation, 2011, p. 59.

of. The dynamics of these discussions cover a great diversity of challenges, ranging from orbital accidents (collisions or even space weather related effects leading to unexpected breakdowns) to intentional threats. Such issues have not been subject to stabilized doctrines and agreements yet and an important part of the solutions will be linked to on-going international discussions. The difficulty of current norm-setting exercises and negotiations on the international scene does not precisely incite to see an easy way forward.

Thus, from a French perceptive, all this does not plead for adopting a "space deterrence"-based approach that would use classic models. On a more modest but maybe more realistic note, the document prefers to refer to the need to "discourage and thwart action by any ill-intentioned third party." It goes on precising that "[military space operations] (...) span various passive and active measures relating to prevention, taking a comprehensive approach (diplomatic, media, economic, legal, etc.); the resilience of all space assets; the defense in space of our space assets," all this being intended to, "preserve freedom of access to and action in space."[251]

VIII -A Policy Based on the Usual Three Pillars of Space Security

Within those main limits (no kinetic energy operations, adoption of practical measures aiming at discouraging attacks), the French posture has been basically relying on the three traditional pillars of existing national space security policies.

The first one, and may be the most important for France in the current context, is the enhancement of its space surveillance (or even space situational awareness) capabilities. The accent has been put on the necessity to own a sufficient capability in this domain, again a direct sign of the need to ensure from A to Z a certain level of autonomy in the use of military space systems. To date, France has modestly but regularly invested in national space surveillance systems. The

251. Space Defense Strategy, p.38.

GRAVES[252] Radar, the main component of the French space surveillance network, was declared operational and put under the Air Force operational control in 2005. This bi-static system is devoted to the monitoring of objects on Low Earth Orbits and has recently been subject to system upgrade in order to preserve its performances for smaller and smaller objects. As announced in the French strategy, a new technical upgrade is expected in 2022 with a "capability increment" planned for 2025. Other means such as the adapted use of a few Air surveillance radars (SATAM radars[253]) or optic monitoring systems for the GEO belt surveillance (including the Airbus-owned Geotracker telescope network) complete the picture. Clearly, this national investment does not meet all the needs that will be required for keeping up with the space traffic development. Many constellations of several or even thousands of small objects are expected on orbit in the coming decades, rendering an effective space picture ever more challenging.[254] Of course the development of new monitoring capabilities has been announced along with increased cooperation. While the French strategy declares that space monitoring capabilities "will have to be subsequently extended to include additional resources in orbit,"[255] it also explicitly plans for example to cooperate with Germany for new satellite radar imaging capabilities.[256] In this respect, the European cooperation is clearly hailed as a necessary way for enhancing national capabilities even if the first

252. The acronym stands for *Grand réseau Adapté à la Veille Spatiale* (Large Network Adapted to Space Survey). For a more descriptive context, see Pasco (Xavier), *A European Approach to Space Security*, American Academy of Art and Sciences, Cambridge, Ma, United States, 2009, pp. 30-35.
253. With a replacement also scheduled for 2030.
254. For Florence Parly, the challenge for the envisioned successor radar amounts to detect "*shoe-box size satellites at a distance of 1500 kilometers*" (Florence Parly speech, 28 July 2018).
255. Space Defense Strategy p.47. Florence Parly has been more explicit by mentioning that the future French military telecommunication satellites will be equipped with surveillance cameras "for their self-protection" and adding that dedicated nano-satellites as well as commercial capabilities are planned to be used to monitor space objects from the orbit (Florence Parly speech, 28 July 2018).
256. Idem, p.46.

attempts in the field of space surveillance have been slow to take shape.[257] In this domain, it is also reminded that the objective of a more European integrated approach "does not rule out the pursuit or development of cooperation with other international partners, especially the United States."[258] As recalled in the strategy, the alliance with the United States remains a key resource in the establishment of a comprehensive space picture.[259]

A renewed technology investment in the passive protection of space systems is also announced as a second pillar of the strategy: "Plans are already in hand for the protection of future defense satellites (Syracuse IV, CERES and CSO). Additional measures must be taken where relevant in order to make them more resilient. In the longer term, technological initiatives will be taken in order to harden the future generation of satellites." In line with the announcement made in the SSA domain, it is recalled that "observation of the close environment of satellites may also help to strengthen their protection. Capabilities of this type will be implemented on board the Syracuse 4A and 4B satellites in order to detect any approach by another satellite."[260]

The acquisition of "a capability to defend our interest," another way of evoking this new "active defense" concept already described, represents of course the last pillar of this trilogy. It finds obvious analogy with the so-called "offensive counter space capabilities" sometimes mentioned in US documents for some years already even if in the French case as already mentioned, the scope may appear clearly

---

257. In 2014, a European Union *Space Surveillance and Tracking* program (EU SST) has been initiated by the European Union under the form of a 5-country consortium (composed by France, Germany, Italy, Spain and the United Kingdom), extended in 2018 to 3 more partners (Poland, Portugal and Romania). To date, some common services have been implemented based on a cooperative use of national capabilities at the EU level. The initial objective of developing new capabilities in this context seems to remain more long term (see the consortium website https://www.eusst.eu/).
258. Space Defense Strategy.
259. Idem, p.34.
260. Idem, p. 48.

limited to the development of carefully selected technologies.

IX –Betting for a New Space Security Initiative

These recent announcements made by France legitimately raise the issue of their compatibility with the country's engagement for a safer, more secure and more sustainable space. Obviously, this willingness to "give the armed forces the capabilities to defend us in space" as expressed in the French strategy[261] may appear as running the risk of opening a Pandora box by sending a wrong signal to the international community. Still, this new policy is meant to be perceived as having simply taken note of a developing situation, as clearly evoked by many administrative experts and operators met by the author for this paper. For sure, the new posture has expressed a somewhat rather disillusioned vision of the way space will evolve in the next future. As described already, along with a perceived growing dependency on space assets, new threats have now been taken for granted, with the space landscape expected to become even fuzzier due to a growing number of space objects which means, for the military, a more hostile and possibly a more dangerous environment they have to prepare for. The publicity surrounding the new space defense strategy also assumes this new pragmatism and as well as a readiness to confront militarily and politically this situation.

French authorities have not given up the parallel objective to actively help establish confidence building measures in space. More than that, while it may not look like it at first glance, part of the strategy can be viewed as a political bet by opening up the debate on the nature of defensive measures that can be taken in space and make it a pragmatic step in the pursuit of a more efficient and realistic collective security in space. In other terms, as can be judged from the political reading of the new doctrine, this military reinforcement can be viable only if it is strictly controlled and legitimate and accompanied by a proactive diplomacy promoting more transparency and confidence-building measures in space. While it may not be easy to

261. Idem.

convince the international community, this is where this new development may still sow useful seeds to possibly unlock most recent international space security discussions.

For many years, the "Prevention of Arms Race in Outer Space" mechanism (known as PAROS) has been the major forum dedicated to space discussions under the auspices of the Disarmament Conference in Geneva. As of today, the international space law consists in five international texts signed in the 60s and 70s, among which the Outer Space Treaty opened for signature in 1967 appears as the corner stone of the entire system. Besides proclaiming space as "the province for all mankind" and "not subject to national appropriation by claim of sovereignty, by means of use or occupation, or by any other means," the treaty does require from the signatories in its article IV "not to place in orbit around the Earth any objects carrying nuclear weapons or any other kinds of weapons of mass destruction, install such weapons on celestial bodies, or station such weapons in outer space in any other manner."[262] While promoting the use of outer space "exclusively for peaceful purposes" in the same article, the State Parties have also quickly agreed on the fact that the text was not intended to prohibit the use of space for defense purposes as those could not be assimilated to non peaceful activities per se. Indeed, many military activities have been conducted since the 1960s, with notably a key role given to surveillance satellite during the Cold War period. Similarly, nothing in the text is prohibiting the test and use of conventional weapon systems.

Given the recent developments in the space domain, with the resumption of antisatellite tests but also with the sky-rocketing development of the orbital population, most of the space faring nations have realized the need for updating this legal international apparatus.

---

262. *Treaty on the Principles Governing the Activities of States in the Exploration and Use of Outer Space, Including the Moon and Other celestial Bodies*, United Nations Office for Outer Space Affairs (text of the Treaty retrieved at https://www.unoosa.org/oosa/en/ourwork/spacelaw/treaties/outerspacetreaty.html).

But beyond this general agreement, the discussions have been basically stalled for more than 10 years with two different approaches fighting against each other. China and Russia have introduced the project of elaborating a new treaty on the "Prevention of the Placement of Weapons in Outer Space, the Threat or use of Force against Outer Space objects" (PPWT) since 2008 and have developed a powerful diplomatic effort to promote this option. This approach calls mainly for prohibiting the placement of "any weapon in outer space" and insists on the legally binding character of the proposed treaty. This text has been largely criticized, especially by the United States but also in Europe on the basis of several perceived difficulties inherent to the text itself. In particular, the difficulty to define the concept of "space weapon" as well as the goal to limit only space-based weapons have often been underscored by the critics. After all they say, most of the recent anti-satellite tests have been performed by ground-based interceptor missiles that would not be covered by the proposed treaty. Additionally, verification would remain a daunting challenge and would render the treaty ineffective for many observers.

On its side, the European Union has introduced in 2008 a project of an "International Code of Conduct for Outer Space" which has consisted in promoting confidence-building measures based on a greater transparency in the conduct of space programs and on spreading the adoption of good behaviors in space. At the difference of the PPWT project, this text was conceived as politically engaging but not legally binding with the assumed objective to facilitate the signature by many State parties and rapidly become an effective platform for policy dialogue. For some, the bet was that such a non binding text could even be viewed as a first inciting tool in diplomatic discussions that would take years for reaching more formal, and more binding agreements, should such opportunities arise.

For more than 10 years now, none of these competing initiatives has succeeded. More particularly, the EU initiated code has suffered a noted blow in 2015 failing to convince the General Assembly of the United Nations (U.N.) and largely dismissed as a possible instrument for conducting space security discussions. It must also be noted that a last parallel effort has been pursued by the U.N. Committee for the Peaceful Use of Outer Space (COPUOS) since 2010 to reach an agreement on "Guidelines for the Long-Term Sustainability of Outer Space Activities". While this activity has produced some encouraging results,[263] the scope of the effort did not include security related issues and all the topics addressed by the Committee were not completely solved.

For many years France has pushed for those discussions and was obviously one of the main promoters of the EU-led initiative. An analysis of the national policy in this regard shows that France has constantly promoted the ICoC as a way to avoid mutually exclusive approaches. On February 11, 2008, in a speech delivered in Kourou (French Guyana), then French president Nicolas Sarkozy insisted on the few principles forming the basis of the French engagement in support of the ICoC project:

> "We must base our action on three principles: freedom of access to space for peaceful purposes, the preservation of the security and integrity of satellites in orbit, respect for the right of states to self-defense. We must start finding pragmatic international solutions to this concrete problem."[264]

For the anecdote, this took place just one day before the proposal for a new PPWT treaty made by China and Russia…

This quest for "pragmatic international solutions," including the right to exert self-defense which has been seen at best controversial, has remained a permanent feature of the French position relative to any international agreement on space security. In a sense, the 2019 policy has shown a great continuity in this respect. It has simply pushed the envelope by acknowledging that solutions would indeed

---

263. The signature in June 2019 of 21 guidelines out of the 33 proposed for discussion has appeared as a notable success after years of discussions. See https://www.unoosa.org/res/oosadoc/data/documents/2019/a/a7420_0_html/V1906077.pdf.

264. Nicolas Sarkozy, 11 February 2008, Kourou (French Guyana). For the complete speech in French, see https://www.vie-publique.fr/discours/169655-declaration-de-m-nicolas-sarkozy-president-de-la-republique-sur-les-d.

have to be found with a space environment already counting active military systems.

X –From "Safety Zones" to "Self-Defense Zones": A Political Threshold?

In first analysis, such a proactive posture moves the lines and creates de facto new room for discussion. After all, protecting satellites also means introducing possible self-defense zones (as suggested by the French doctrine), and this immediately raises the issue of the right to intervene while no territory can be claimed as sovereign in space. Indeed such a "territorialization" of space cannot be conceived in the current legal context. This is precisely where this deliberate move to inject a high degree of political realism seems considered by itself as a possible way forward that would hopefully help overcome the stalemate in the ongoing discussions. Obviously, this deliberate positioning is on a tightrope and may likely stir strongly negative reactions. But in an ideal scenario, it may also have the merit to help reconnect those discussions with "real-life" situations. This obviously implies a great degree of transparency. Ensuring that such discussions can take place means being clear beforehand about one's own intentions and not using pretenses. The choice that is being made to clearly distinguish between the military actions to be proscribed (in this case kinetic energy weapons) and those acceptable based on controlled means and effects can then become a legitimate subject for discussion.

Interestingly enough, the dynamic in favor of discussions on the existence of self-defense zones could benefit from new, controversial developments, a priori far removed from orbital defense concerns. On May 15, NASA Administrator Jim Bridenstine publicly presented the ARTEMIS Accords, an initiative to develop international standards of behavior for the future lunar exploration program. If they reiterate the main principles of traditional space law (peaceful purposes, interoperability, emergency assistance, registration of space objects, release of scientific data, access and use of space resources), these future agreements under discussion also include new features, one of which deals with the "Deconfliction of Activities." And in this context, the mention of "safety zones" appears for the first time. This mention has obviously not failed to provoke a negative reaction in many countries, particularly in Russia, where some commentators have seen it as a harmful drift towards the constitution of sovereign zones in space and on celestial bodies. However, in the presentation of what can pass for a legal innovation, NASA seems to be simply appealing to the responsibility of those involved in exploration. "Specifically, via the Artemis Accords, NASA and partner nations will provide public information regarding the location and general nature of operations which will inform the scale and scope of 'Safety Zones.' Notification and coordination between partner nations to respect such safety zones will prevent harmful interference, implementing Article IX of the Outer Space Treaty and reinforcing the principle of due regard."[265] In fact, this particular initiative can be seen as deriving directly from the spirit of the article cited by the US space agency. It simply requires the signatory states to conduct their activities "with due regard to the corresponding interests of all other States Parties to the Treaty."[266]

For all that, and even if the Artemis Agreements do not speak of a prohibition of access, the simple prospect of "safety zones" on the Moon raises the question of a certain level of access control and activities on perimeters defined by national activities and their compatibility with Article II of the OST. Indeed, the latter clearly states that "Outer space, including the moon and other celestial bodies, is not subject to national appropriation by claim of sovereignty, by means of use or occupation, or by any other means."[267] In fact, the treaty even goes so far as to reinforce this spirit of openness (in the literal sense) through its Article XII, which induces the positive obligation to keep any facility accessible to possible visits.

These Artemis Agreements are still under discussion among a lim-

265. https://www.nasa.gov/specials/artemis-accords/index.html.
266. https://www.unoosa.org/oosa/en/ourwork/spacelaw/treaties/outer-spacetreaty.html.
267. Idem.

ited number of space agencies representing the former space station partner countries and as mentioned, the proposal initially aroused great mistrust and even opposition from major space countries such as Russia and perhaps China (although these positions may not be definitive). The fact remains that this type of agreement symbolizes an unavoidable confrontation between the evolution of space technologies and the legal texts that frame their use. And in a certain way, these debates will necessarily extend to practices in orbit, if only to allow the repair or refueling of satellites. These maneuvers will involve regulating satellite approaches and will also require a certain degree of transparency. Ultimately, all this will obviously concern the military.

De facto, the French vision of 360° surveillance operated from "patroller satellites," with the possible implementation, repeated in 2020, of laser systems or jammers if necessary, also brings into play this notion of "safety zones" aimed at fighting against any "harmful interference."[268] But the French discourse suggests the possibility of rules of engagement based on the existence of exclusion zones, which in itself would constitute a real first from the point of view of international law.[269]

On this point, and unlike the Artemis Accords, the resources that can be found in the interpretation of existing texts appear rather limited here. No possibility is given in the treaties for the establishment of no-access zones around satellites to be protected. If one adopts the French vision, it is therefore first of all a new political deal that the countries would choose to agree on and that would form the basis for the collective adoption of new norms of behavior in defense mat-

---

268. Cette notion est également évoquée par la NASA en appui de son argumentaire en faveur de " safety zones " pour la Lune (https://www.nasa.gov/specials/artemis-accords/index.html).

269. Il faut noter que les discussions précurseurs au débat sur la Space Force américaine ont mis le thème de rules of engagement in space à l'avant-scène et les présentent comme un besoin devenu essentiel. See for example U.S. Military Lacks Clear Rules of Engagement in Space, National Defense, 22 March 2017 (https://www.nationaldefensemagazine.org/articles/2017/3/22/us-military-lacks-clear-rules-of-engagement-in-space).

ters. Given the international community's difficulties in agreeing on shared principles of space security, discussions on more sensitive subjects have a priori little chance of succeeding (or even starting).

Then, can a scenario be imagined in which all subjects could be addressed? It is far from sure. However, some first steps can be envisioned. A first necessary effort would be to reach a common vision of the multiplication of space operations and the need to regulate these new operations. As mentioned earlier, discussions exist today in various forums and they can be the basis for further study. In particular, the topic of Space Traffic Management (STM) has become one of the major discussion points for regulating maneuvers in orbit. STM could provide a framework for initial reflections focused on military developments.

Depending on the progress of the debates on the STM, a few criteria can also be imagined that distinguish military activity and identify its specific protection needs. This would represent the second necessary step in building a common view. Several instruments of different natures already exist that can inspire further discussions. For example, the Hague Code of Conduct against the proliferation of ballistic missiles, introduced in 2002 and signed by 140 countries since then, has introduced the notion of launch pre-notifications among its recommendations in order to avoid any misunderstanding about missile testing and space launches in the international community. In this highly sensitive and politically heavy-loaded domain, such common practices have been accepted and mostly applied, even if with a relatively modest impact. It may not be completely unrealistic to envision similar pre-notifications dealing with orbital objects maneuvers that could be part of renewed discussions on commonly acceptable regulations. First progresses that could be made in this respect may then help further discussions regarding maneuvers in orbit more generally. The possibility to hold preliminary discussion on criteria such as distances and proximity between satellites, the level of transparency regarding the function, the activity, the status, or the refreshed orbital data of a satellite when maneuvered may naturally lead to define de facto "safety" or protected zones in an acceptable way. Put in another way, the possibility to establish such zones, with perimeters ranging say from dozens of kilometers to hundreds of meters, may then only make sense if a more general political agreement on definitions of good behavior in orbit have been agreed upon in the first place. This would obviously imply technical discussions but after all, precedent exists with the Interagency Debris Commission (IADC), a gathering of some 15 space agencies, that has discussed for decades common agreements on the management of space debris. Such a model can again certainly apply as soon as Space Traffic Management can be discussed, encapsulating security issues.

Of course, such developments may seem hypothetical and remote. But the high pace of change in space activity has put the safety and security of in-orbit operations back on the radar screen faster than expected. In this context, the military dimension inherent to these new international discussions and the interest of "safety zones" in space take on their full meaning. One must no doubt see in the French discourse a willingness to be part of this dynamic by trying to avoid being out of time; at the risk of brutal reactions that may reject the very spirit of this position, which remains essentially national for the time being.

Whatever its objective, the new French doctrine cannot avoid a revival of space diplomacy. This aspect was indeed often cited by the Minister as an essential effort. But even more so, this revival may appear to be a sine-qua-non condition for ensuring its political credibility. France, more than others, will have to multiply initiatives in this field. This real-world based approach of concerted security in space does go with political consequences which, if well managed, could relaunch the strategic space dialogue on new bases.

XI –Conclusion: Towards New Norms for Space Military Activities?

At the end of this analysis, the new French Space Defense Strategy, beyond its military meaning, can also be viewed as political opportunity that has been seized to deliberately address the complex realities of space security.

The message has been clearly sent out: given the new confrontation on the international scene, France considers that the current dynamics in space, involving possible showdowns as well as a rise of unpredictable events in space cannot be curbed. This fact must be faced it is proclaimed, both nationally and internationally. Many new efforts have started to install new rules of the road and Space Traffic Management or collaborative Space Situational Awareness sound as familiar expressions in the ears of most space policy specialists. On a different note, a recent paper has argued in favor of building a space coalition to join forces and make sure it can win a space war. The author asserts that "Washington would be wise to build a coalition that creates a more robust system for monitoring the space domain, reacting to space threats, and prevailing in a war that extends into outer space. Building a space coalition — especially one that includes allies from Europe, the Middle East, and Asia — is no simple task. But, working towards system integration and also a common space doctrine is positive step forward towards increased resiliency of the space systems that are essential for precision warfighting." But the author also recognize that diplomatic efforts should not be neglected as well. Of course, technical systems cannot be the only focus of the coalition; it should also devote attention to the development of norms for military space activity."[270] However, while the ideas are there, the means to implement them and their governance divide the international community and are not sufficient to moderate military ambitions. For sure, time has come for such discussions to be held internationally but they cannot be reserved to the traditional competing of major powers.

In this context, the new French doctrine announces the strengthening of the national military while caring not to dismiss the value of a continuous genuine diplomatic effort. The announcement of technological developments, whether for laser systems or in the field of SSA is perfectly reflecting the dual nature of the proposed approach. SSA systems perfectly represent the cooperative spirit and transpar-

270. Bateman (Aaron), *America Needs a Coalition to Win a Space War*, War on the Rocks, April 29, 2020; See https://warontherocks.com/2020/04/america-needs-a-coalition-to-win-a-space-war/.

ency that space needs[271] (and it is recalled in the strategy), while the laser system signals a preferred national option for self-defense. Those are complementary steps for a reasoned space defense doctrine that does not have to jeopardize international agreement as has been recalled by the minister in a rather "right-on-target" style.

Some critics may be fair in pointing out at apparent contradictions. It may be equally fair to consider that moving the terms of the space security debate from the complete prohibition of space weapons to the prohibition of certain types of weapons may help call for an international ban on the most destructive ones. This position, supported by the new French strategy, may bring new meaning to a debate that has been completely locked up for many years.

This does not go without political difficulties and responsibilities. France has been one of the most prominent supporters of the EU–led project of and International Code of Conduct for space, calling for more transparency and confidence building measures. It will certainly have to do more to convince the international community of the necessity of a new start that can satisfy the two ends of the space negotiation spectrum. In this respect, accompanying or complementary policies, for example in actively promoting rules for a better Space Traffic Management, will be necessary to defuse any misunderstanding about the intent of future space security discussions.

Promoting an approach that legitimates self-defense measures and can guarantee that they can be implemented in a way that complies with controlled norms of behavior is certainly a huge challenge. This is after all about creating new rules for the security of each nation in space that would include "acceptable" defensive systems, ultimately leading to the strengthening of the space dialogue between nations.

Such a scenario may barely look within reach. This is an external observer hope. But it takes into account the real state of the national postures that we see developing in space and which are unambiguously engaged in the over-militarization of space. While possibly

271. A number of international agreements already exist to organize data sharing procedure in this domain.

contributing to reinforcing this trend, the French posture may also be seen as taking care to prepare for future discussions and propose ways to prevent an uncontrolled arms race in space. At least, the reader can agree that this is a hope that we cannot afford to neglect.

Chapter 5:

Countering Co-Orbital ASATs: Warning Zones in GEO as a Lawful Trigger for Self-Defense

David Bernstein. Michael B. Cerny, Brandon W. Kelley,[272] Raphael J. Piliero

Abstract

The existing literature on space arms control is plagued by a notable gap. Despite widespread acknowledgment of the potential threat posed by co-orbital counterspace operations, particularly in geosynchronous orbit (GEO), one of the most promising solutions remains undertheorized: the declaration of safety zones around satellites. After four decades of debate, proponents and critics have persistently failed to synthesize arguments for and against such proposals in a comprehensive and systematic fashion. The result has been predictable: critics have given unduly short shrift to such proposals, while advocates have failed to sufficiently address their greatest shortcomings. One prominent advocate recently noted regarding a critic of zone proposals that "we agree on all the facts, yet our conclusions are often opposite". This paper demonstrates that both sides of the debate have merit, and that there is a path to agreement that navigates these contradictions. Critics are correct that, as

272. Corresponding author: Brandon W. Kelley, Associate Director of Debate at Georgetown University. Center for Security Studies, Georgetown School of Foreign Service. 3700 O St., N.W. Washington, D.C. 20057. Email: bwk9@georgetown.edu

currently framed, most zone proposals violate international law or are undesirable on policy grounds. However, this need not be true. Framed solely as an information-gathering mechanism supporting the conditional and preemptive use of force in self-defense, "warning zones"—distinguished herein from both "self-defense zones" (SDZs) and "keep-out zones" (KOZs)—are a lawful and effective solution to co-orbital anti-satellite (ASAT) threats in GEO. Designed properly, such warning zones also avoid the most salient policy objections to SDZs and KOZs, while constituting the only presently effective means of addressing a potentially catastrophic threat to space security. They also offer a rare opportunity for truly cross-ideological agreement, and a way forward for international space arms control discussions—something sorely needed in this current political moment.

Introduction

There is increasing realization among both space policy experts and government officials about the vulnerability of United States satellites to attack or disruption by other states. In recent years, one such threat is co-orbital counterspace operations (also known as rendezvous and proximity operations, or RPOs).[273] Under such a scenario, a United States satellite could be closely approached or tailgated by the satellite of another entity in order to disable or destroy it. This threat is particularly pressing, as the "dual-use" nature of activities in space allows an aggressor to masquerade as a civilian satellite under an ostensibly peaceful purpose such as active debris removal or on-orbit servicing.[274] Due to the

273. Chow, Brian G., and Brian Weeden. 2020. "Transcript of a Discussion between Dr. Brian Chow and Dr. Brian Weeden on Space Zones and Bodyguards for Proximity Operations." Presented at the Panel II of a Space Policy Workshop Co-sponsored by the Nonproliferation Policy Education Center and the American Bar Association Standing Committee on Law and National Security. http://npolicy.org/article.php?aid=1465&tid=2.
274. Ibid.

now burgeoning space industry, experts worry that the increase in government and civilian space activities would increase the vulnerability of United States satellites.

Open-source reporting has confirmed that the threat posed by co-orbital operations is growing. A 2020 report by the Secure World Foundation evidence that the Russian Federation is actively testing advanced RPO technologies in both low-earth orbit (LEO) and geosynchronous orbit (GEO).[275] At least two active programs[276] are ostensibly non-aggressive, but the "dual-use" nature of the technology could nonetheless be applied in military or strategic contexts. Similarly, the People's Republic of China may be close to possessing such capability; while China has not yet conducted a destructive co-orbital intercept of a satellite, China has conducted multiple RPO tests in both LEO and GEO.[277] Highlighting concerns about the nature of this technology, a report by the Center for Strategic and International Studies also concluded that both Russia and China are testing the ability to maneuver satellites around one another.[278]

Open-source intelligence also confirms many concerns about the threat posed by co-orbital counterspace technology. A new Russian program, *Burevestnik,* in Russia's Central Scientific Research Institute of Chemistry and Mechanics (CNIIHM), referred to as a "space security complex" in official documentation, is believed to be developing the use of microsatellites to target enemy satellites in GEO.[279] This fact is unsurprising considering Russia's long history

---

275. *Global Counterspace Capabilities: An Open Source Assessment*. 2020. Washington, DC: Secure World Foundation. Pp XII. https://swfound.org/media/206970/swf_counterspace2020_electronic_final.pdf.
276. The Secure World Foundation lists a co-orbital ASAT program referred to as *Burevestnik* and a surveillance and tracking program called *Nivelir*.
277. *Global Counterspace Capabilities: An Open Source Assessment*. 2020.
278. Harrison, Todd et al. 2020. *Space Threat Assessment 2020*. Washington, DC: Center for Strategic & International Studies. https://aerospace.csis.org/wp-content/uploads/2020/03/Harrison_SpaceThreatAssessment20_WEB_FINAL-min.pdf.
279. Hendrickx, Bart. "Russia's secret satellite builder," The Space Review. May 6, 2019. https://www.thespacereview.com/article/3709/1

of co-orbital technology development. As early as 1961, the Soviet Union attempted to develop an early co-orbital system (the Istrebitel Sputnikov), and in the 1980s, the Soviet Union attempted to create a higher-altitude co-orbital system known as the Naryad.[280] Recent observations of close Russian satellite approaches (such as the close approach of the USA-245 in 2020) suggest a resurgence in Russian co-orbital interest.[281]

Co-orbital capabilities represent a significant threat to the United States most critical satellites. Given American reliance on space for all facets of military operations, the risk is particularly acute. Unlike during the Cold War, the contemporary US military relies on satellites for its "eyes and ears," providing missile-defense warnings, battlefield communications, and signals for guiding precision munitions.[282] In GEO, the United States maintains high-orbit satellites to alert the military when other countries launch ballistic missiles, and these "early warning" satellites serve as the lynchpin of security.[283] In a scenario described by Dr. Brian Chow, close approaches could be used to pre-position multiple enemy satellites near critical United States early warning satellites.[284] With the ability to attack at a moment's notice, the United States satellites are currently defenseless to this form of attack.[285] Furthermore, the compressed response time is what makes co-orbital attacks uniquely challenging to counter. While a direct-ascent ASAT attack in GEO would

280. Thompson, Cort. 2020. "Russia's Recent Space Activity Is a Return to Old Form." *Lawfare*. https://www.lawfareblog.com/russias-recent-space-activity-return-old-form (May 25, 2020).
281. Ibid
282. Davenport, Christian. 2019. "The Battlefield 22,000 Miles above Earth." *The Wilson Quarterly*. https://www.wilsonquarterly.com/quarterly/the-new-landscape-in-space/the-battlefield-22-000-miles-above-earth/ (May 25, 2020).
283. Arbatov, Alexey. "Arms control in outer space: The Russian angle, and a possible way forward," *Bulletin of the Atomic Scientists*. Vol, 75, No. 4. 2019
284. Chow, Brian G., "Stalkers in Space: Defeating the Threat," Strategic Studies Quarterly 11, no. 2 (Summer 2017): 82, https://www.airuniversity.af.edu/Portals/10/SSQ/documents/Volume-11_Issue-2/Chow.pdf.
285. Chow, "Stalkers in Space," 82.

give the United States only four to six hours to both anticipate and respond, a pre-positioning co-orbital attack would be comparatively instantaneous.[286]

Such risks are also not theoretical. Many incentives exist for Russia and China to press United States boundaries in space and conduct close approaches. According to the "Senkaku Paradox" articulated by Michael O'Hanlon, China might start with small acts of aggression in the South China Sea or Senkaku Islands to test American resolve.[287] Although it is unlikely China would engage in overt military attack, these low-level acts of aggression represent gambles over an "essentially worthless collection of rocks."[288] According to O'Hanlon, such attacks would likely begin in space due to the United States' vulnerability, as destruction or disruption of critical satellites could hamper a timely and effective American response. Furthermore, a public narrative that China or Russia could effectively deny American responses by disrupting critical satellites poses a latent risk to American alliances and assurances, even if such attacks never occur.[289]

Independent of the risks to American security commitments, the space domain itself is also prone to misunderstandings and miscalculation. Space is an "offense-dominant" domain, where satellites are inherently vulnerable and difficult to defend, lending an advantage to the attacker. This creates a structural incentive for a state with substantial dependence on space to strike while the iron is hot, protecting their own satellites.[290] The combination of compressed

---

286. Swarts, Phillip "Space Wars: The Air Force Awakens," *Air Force Times*, 2016, available at https://www.airforcetimes.com/news/your-air-force/2016/02/15/space-wars-the-air-force-awakens/.
287. O'Hanlon, Michael "Can America Still Protect Its Allies? How to Make Deterrence Work," *Foreign Affairs*, Vol. 98, No. 5. September/October 2019.
288. Ibid.
289. See, for example: Brandon J. Weichert, "Preparing for a Russian 'Space Pearl Harbor,'" Orbis, June 14, 2019, https://doi.org/10.1016/j.orbis.2019.05.002.
290. Grego, Dr. Laura, Union of Concerned Scientists, Space and Crisis Stabil-

timelines for responses with a lack of crisis management history create a lethal cocktail for a dispute to spiral out of control, with the ensuing conflict spreading to the terrestrial domain.[291] Deterrence also runs unique crisis stability-related risks in space, as it requires states to consistently demonstrate capability and resolve;[292] the aforementioned characterization of space as an offense-dominant domain makes this uniquely unstable.

However, some experts disagree with the extent of the threat posed by these operations.[293] Performing a close approach to an enemy satellite would present a challenge, let alone successfully disabling an enemy satellite while remaining undetected. Instead, an aggressor would likely resort to any number of other options, such as a terrestrially launched ASAT attack, signal jamming, or a cyber-attack.[294] Yet, concerns persist within the United States federal government and the United Nations about the threat posed by co-orbital capabilities. In 2018, US Vice President Mike Pence commented that "Both China and Russia have been conducting highly sophisticated on-orbit activities that could enable them to maneuver their satellites into close proximity of ours, posing unprecedented new dangers to our space systems."[295] These concerns have been echoed by working groups of the United Nations, and, in 2019, the UN Disarmament Commission examined the threat posed by co-orbital satellite operations.[296]

---

ity, March 2018, https://www.law.upenn.edu/live/files/7804-grego-space-and-crisis-stabilitypdf.

291. Ibid.

292. Ibid.

293. Chow, Brian G., and Brian Weeden. 2020. "Transcript of a Discussion between Dr. Brian Chow and Dr. Brian Weeden on Space Zones and Bodyguards for Proximity Operations." Presented at the Panel II of a Space Policy Workshop Co-sponsored by the Nonproliferation Policy Education Center and the American Bar Association Standing Committee on Law and National Security. http://npolicy.org/article.php?aid=1465&tid=2.

294. Ibid.

295. Remarks by Vice President Pence on the Future of the U.S. Military in Space. August 9, 2018. http://spaceref.com/news/viewpr.html?pid=52955.

296. UNDIR. "Practical Implementation of Transparency and Confidence-

To address the threat posed by these operations, experts and government contractors have proposed that the United States establish a policy by which zones would be designated around American space assets, which, if breached, would justify an armed response against the potentially aggressive satellite.[297] Although significant disagreement persists about the legal, technical, and political feasibility of these zone proposals, there exists common ground amongst proponents and critics about the value of signaling state intent in outer space. United States military officials have acknowledged that, even if space zones are informal, they provide a means of registering a grievance; another official noted that zones can help communicate a menu of increasingly escalatory options in space, similar to the use of flares or warning shots for maritime vessels.[298]

Furthermore, the United States has a unique opportunity to be the standard-bearer for a new form of arms control to address co-orbital counterspace technologies. This is intentionally differentiated from a strategy based primarily on deterrence in space. A deterrence-driven approach would, in the words of Professor David A. Koplow, be "expansive and therefore expensive," requiring the US to be perpetually on the look-out for any potential threat in space.[299] In contrast, arms control could be mutually agreeable and effective. Professor Koplow argues that ASATs are nascent enough to be effectively regulated while in the development and testing phase; ASATs have yet to be used in conflict, which makes now a particularly opportune time for a mutual accord.[300] The US has an interest in protecting the

---

Building Measures (TCBMs)" Briefing Series for UN Disarmament Commission Working Group II. April 10-12, 2019.

297. Ibid.

298. American Bar Association Standing Committee on Law and National Security, "Working Smarter with America's Spacefaring Allies," Workshop Report. March 2020. http://npolicy.org/article_file/march-2020-space-report.pdf

299. David A. Koplow, Deterrence as the MacGuffin: The Case for Arms Control in Outer Space, Updated January 2020, 10 J. NAT'L SECURITY L. & POL'Y __ (forthcoming), https://jnslp.com/wp-content/uploads/2020/01/Deterrence_as_the_MacGuffin.pdf pp 38.

300. Ibid, 72.

space capabilities that it relies on for war; meanwhile, China and Russia have an interest in enshrining binding restraints on further advances that the US could make with regards to military space capabilities.[301]

Indeed, a failure to actively push for arms control creates a concomitant risk to US credibility. While China, Russia, and the European Union have advanced their own series of arms control proposals, the US has mostly remained on the sidelines. Douglas Loverro, the former Deputy Assistant Secretary of Defense for Space Policy, has argued that US must take ambitious actions to reverse a reputation of intransigence. He argues that "[a] clear consensus arose that whatever reasons the United States may have for not being willing to negotiate a binding treaty like the PPWT or a non-binding instrument like the ICoC, it should propose alternative measures which it is willing to support. Rebuffing international attempts to create more collaboration without providing alternatives means that the United States is failing to take a leadership role."[302] Even skeptics of the zone proposal agree, with Dr. Weeden noting that " US leadership has waned… space security issues… the lack of strong US engagement… allowed Russia, with… China to… seize the diplomatic narrative and ultimately derail the negotiations... the US has long refused to offer any substantive proposals, leaving Russia and China to set the agenda with their proposals for banning space weapons and no first placement of weapons in space, both… designed to provide advantages… over the US." [303] Furthermore, Weeden writes that the US could, "propose restrictions on the ability of satellites to engage in close approach… thereby clarifying threatening behaviors and enabling the use of self-defense measures…. rallying allies

301. Ibid.
302. Loverro, Douglas, "Summary of Discussions of the Expert Roundtable on The Weaponization of Outer Space: Ethical and Legal Boundaries," *Center for Ethics and the Rule of Law.* April 6-7, 2018. Pp 24. https://www.law.upenn.edu/live/files/10006-cerl-conference-summary-report-weaponization.
303. Weeden, Brian. "The Space Review: The Trump Administration Needs to Exercise Leadership in Space Security Diplomacy," August 20, 2018, http://www.thespacereview.com/article/3554/1.

and partners to our call..."[304] One potential proposal would be warning zones in space.

The following report synthesizes historical and contemporary political, technical, and legal discussions on establishing specialized zones in outer space to protect American satellites from co-orbital ASAT technology. Developing a taxonomy of different zone proposals across the space security and arms control literature, we assess the relative legal merits of these proposals. Our findings suggest that previous proposals would be untenable due to legal limitations on space activities set forth by the Outer Space Treaty. However, we propose the establishment of a new type of zone— 'warning zones' —which, if framed solely as an information-gathering mechanism supporting the conditional and preemptive use of force in self-defense, would constitute a lawful and effective solution to co-orbital ASAT threats in GEO.

Summary of Key Findings

We find the following:

- Preemptive uses of force in self-defense in space are clearly lawful, so long as all relevant proscriptions on its exercise are observed.
- Prior zone proposals, which envision enforcement of zones *per se*, are almost certainly in violation of the Outer Space Treaty (OST)'s prohibitions on national appropriation by means of sovereignty or use (Art. II) and harmful interference with the activities of other States (Art. IX). This paper's proposed "warning zones," in contrast, fully comply with both provisions, placing a subset of safety zones on unassailable legal footing.
- The existing literature regarding legality of zones uniformly fails to consider such a formulation. Proponents have thus far

304. Ibid.

leaned on strained analogies to the ITU, ISS, or law of the sea, which critics have rightly identified as spurious.

- The unique physical characteristics of GEO render all available alternatives incapable of adequately addressing the threat posed by co-orbital interception from prepositioned assets, should technical capacity be(come) sufficient to enable such loitering and interception to be undertaken reliably. This is a realistic possibility, and the resultant risks to international stability could be substantial.

- Both critics and proponents of zones have failed to systematically address each other's arguments. Each side is correct in key respects often glossed over by the other. Designed properly, warning zones can avoid nearly all major objections raised by critics, while meaningfully increasing crisis stability and decreasing the risk of interstate conflict.

- Most saliently: prepositioning of co-orbital assets poses a unique threat undervalued by critics, and addressable in the near-term only by drones. Critics' well-founded concerns that drawing redlines will create perverse incentives to test and increase the propensity for crisis escalation are resolved by this paper's reformulation of zones as a flexible mechanism for deriving signals of possible intent. Meanwhile, claims that the adversary will simply loiter outside the bounds of the zone, or that the zone cannot reliably extend far enough to prevent close-approaches, fail to understand the true value of the proposal, which is *informational*, not prohibitive. Finally, the ability to operate on the edge is a bug, not a feature, as it increases the ability and likelihood of states sending clear signals and adds important rungs of bargaining, short of the use of force, to the escalation ladder.

- Warning zones are a form of space arms control that is appealing to all parties. As a uniquely self-executing crisis stability measure, zones are devoid of the verification concerns, definitional debates, and fears of restricting freedom of action that have brought space arms control negotiations to a grinding halt. This provides a unique opportunity to forge consensus, and take

> the first steps toward a new space arms control agenda. Doves, hawks; Democrats, Republicans; Americans, Chinese, Russians, Europeans: all are well-served—and none are harmed—by increased clarity around intent in orbit.

This paper will begin by taxonomizing space security zone proposals and assessing their (il)legality. Only warning zones (WZs) are conclusively lawful under the OST, and international law places several additional constraints on their operation. The paper will then evaluate the case for WZs on normative policy grounds, comprehensively surveying the arguments for and against such zones. This paper finds that critics' arguments are well-founded, and may very well be damning to other zone proposals; but that in one specific context, WZs with certain specific design elements are capable of resolving such criticisms while providing an invaluable contribution to space security.

Typology of Zones

The dearth of international law regulating co-orbital anti-satellite technology (ASATs) leaves states with few options to combat the burgeoning threat posed by these weapons. Nonetheless, steps must be taken to regulate dual-use space systems before they are deployed *en masse*. A significant number of proposals argue that safety zones around satellites should be established, though the existing literature on these zones elides important legal and policy distinctions between them. This section seeks to rectify the conflation between different types of zones by sketching a clearer zone taxonomy, and then evaluating their relative legal and policy merits. For purposes of this paper, we only consider zone types related to arms control proposals; a more expansive typology of space zones which elaborates on other varieties such as non-interference zones[305] and space traffic management/collision-minimization zones should be subject

305. Cody Knipfer, "The Space Review: Revisiting "Non-Interference Zones" in Outer Space," The Space Review, 2018, https://www.thespacereview.com/article/3418/1.

to further research.

Proposals to establish zones in outer space within which certain conduct by states could be restricted or prohibited date to the earliest days of space arms control during the Cold War.[306] For example, in 1980, *M.I. Lazarev* proposed the establishment of security zones around settlements in outer space through which other states could only pass through peacefully.[307] In 1986, *P.Q. Collins* and *T.W. Williams* proposed a "Traffic Zones" system whereby international law could proscribe certain movements and types of passage by satellites in bands of space reserved for solar power generation and space stations[308] Furthermore, T. K. *Schwetje* identified a variety of other proposals in the 1980s to establish zones designed to protect states' space assets which, if penetrated without permission, would allow retaliation through the use of force.[309]

These proposals may be broadly divided into three categories: 'keep-out zones' (KOZs), 'self-defense zones' (SDZs), and 'warning zones.' This taxonomy is a relatively novel contribution. Works to date use these (and many other) terms imprecisely—for many authors they are interchangeable, while others seek to distinguish their own proposal but fail to provide any systematic account of the legal and policy distinctions among types of zones.

I. Keep-Out Zones: KOZs seek to restrict close approaches by declaring areas of space around sensitive space assets through which unauthorized entry or passage may be met with military force. A

---

306. Schwetje, Kenneth. 1987. "Protecting Space Assets: A Legal Analysis of 'Keep-Out Zones.'" *Journal of Space Law* 15(1 & 2): 131–46.
307. Lazarev, *Future Space Cities*, 5 ANNALS AIR & SPACE L. 529 (1980) in Schwetje, "Protecting Space Assets: A Legal Analysis of 'Keep-Out Zones,'" 132.
308. Patrick Q. Collins and Trevor W. Williams, "Towards Traffic Control Systems for Near-Earth Space," in Proc. 29th Colloquium on the Law of Outer Space (IISL, Space Future, 1986), 161–70, http://www.spacefuture.com/archive/towards_traffic_control_systems_for_near_earth_space.shtml.
309. Schwetje, "Protecting Space Assets," 131–46; Richard Dalbello, "Rules of the Road: Legal Measures to Strengthen the Peaceful Uses of Outer Space" (1985) 28 in Schwetje, "Protecting Space Assets," 132.

1985 report by the United States Office of Technology Assessment (OTA) proposed the archetypal example of a KOZ, which OTA defined as "a volume around a space asset, off-limits to parties not owners of the asset."[310] The OTA noted that such zones "could be negotiated or unilaterally declared," and contended that "the right to defend such a zone by force and the legality of unilaterally declared zones under the OST remain to be determined."[311] OTA's proposed KOZ would ensure that satellites "keep 100 km and 3 degrees out-of-plane from foreign satellites below 5,000 km [and] keep 500 km from foreign satellites above 5,000 km except those within 500 km of [GEO]," would allow "one pre-announced close approach at a time," and would establish that "in the event of… violation… the nation… which most recently initiated a maneuver 'burn' is at fault and… satellites trespassing… may be forcibly prevented from continued trespass."[312] According to the OTA's assessment, such zones would be useful in the defense against passive anti-satellite weaponry.[313] Passive anti-satellite weaponry (ASATs), such as space mines, might loiter near a satellite prior to performing an aggressive maneuver. [314] Furthermore, a principal finding of the report is that the United States could unilaterally "declare and defend protective [KOZs] around critical satellites," which could entail the use of force.[315]

II. Space Defense Zones/Self-Defense Zones: SDZs seek to apportion specific orbital regions for primary use by a particular state or group of states. Unauthorized trespass may be met with force, typically justified on self-defense grounds, but peaceful passage by foreign satellites is allowed with permission. Perhaps the most comprehensive and best-known SDZ proposal comes from Albert Wohlstetter and Brian G. Chow in 1986,[316] and was included in

310. "Anti-Satellite Weapons, Countermeasures, and Arms Control" (Washington: U.S. Congress, Office of Technology Assessment, September 1985), ix.
311. Ibid.
312. Ibid, 136.
313. Ibid, 10.
314. Ibid., x, 9-18.
315. Ibid., 8.
316. Albert Wohlstetter and Brian G. Chow, *Self-Defense Zones in Space* (Ma-

the Commission on Integrated Long-Term Strategy's Congressional report on "Discriminate Deterrence" (1988) as a means for the United States to remedy its weakness in space.[317] Wohlstetter and Chow proposed that the United States and then-Soviet Union establish large, protective sectors in geosynchronous orbit through which foreign satellites could pass peacefully (only) with permission;[318] should a foreign satellite trespass without such permission, there would be an "explicit right to inspect, expel or render harmless invading satellites."[319] There are crucial differences between SDZs and KOZs. First, as Schwetje notes, "rather than attempting to protect every satellite with a critical mission" as do KOZs, "an SDZ protects the mission itself... [and is] not attached to the satellites"; to accomplish this, "Wohlstetter and Chow reserve entire sectors for satellites of various defense groupings... to be assigned to the Western Alliance or the Soviet Bloc."[320] This distinction greatly impacts the threat posed by loitering of co-orbital ASATs, as the "SDZ arrangement would make the entire 'shell' inaccessible to an attacker" such that the "threat of space mines would be lessened."[321] Second, the express articulation of a legal right grounded in self-defense to preempt an approaching threat with whatever force might be required is thought to render SDZ more effective at dealing with violations.

Such proposals are not limited to Wohlstetter and Chow. Dan Quayle, for instance, proposed similar "space self-defense zones" in 1986, writing that they would:

> "... allow defenses, enhance verification, and significantly reduce the incentive for either side to attack the other's sat-

---

rina del Rey, CA: Pan Heuristics, 1986).

317. Albert Wohlstetter and Brian G. Chow, "Self-Defense Zones in Space," Contract, Integrated Long-Term Defense Strategy (Marina Del Rey, CA: Pan Heuristics R&D Associates, July 1986); Fred Iklé and Albert Wohlstetter, "Discriminate Deterrence: Report of the Commission on Integrated Long-Term Strategy" (Comission on the Integrated Long-Term Strategy, January 1988), 54.

318. Schwetje, "Protecting Space Assets," 133.

319. Ibid., 133.

320. Ibid., 133.

321. Ibid., 133.

> ellites. What the zones would do is simple: Keep one side's satellites far enough apart from the other's to make an instantaneous surprise attack by either side virtually impossible. Both sides would be permitted an equal number of zones within which they would have exclusive rights to position their satellites permanently, and to inspect, expel, or otherwise render harmless intruders if they exceed an agreed upon, safe number."[322]

LTC A. J. Butler recommends a layered proposal, in which,

> "...two distinct types of zones would be established. The first would be a true safety zone that required permission prior to entry and would be of limited dimension. The extent of the safety zone would only be as large as necessary to protect activities in which safety was a prime factor. Outside the safety zone, a Space Defense Identification Zone (SPADIZ) would be established. The SPADIZ would be quite large; all approaching spacecraft would be required to identify themselves and give prior notice of approach."[323]

Many more contemporary formulations of the concept retain the central emphasis on self-defense, but alter the details to account for the dramatic evolution in the space environment since the 1980s—the exponential increase in traffic and use and the post-Cold War shift away from bipolarity, among others. Perhaps the most comprehensive proposals in recent years have come from Chow, who now argues for the establishment of SDZs around certain satellites the penetration of which would trigger the right to "punish infringement," at least under certain specified threshold conditions.[324]

---

322. "Dan Quayle, "Making War Less Attractive: Space Self-Defense Zones," *Christian Science Monitor*, March 24, 1986, https://www.csmonitor.com/1986/0324/espace.html.
323. Schwetje, "Protecting Space Assets," 133.
324. See Brian G. Chow, "Stalkers in Space: Defeating the Threat," *Strategic Studies Quarterly* 11, no. 2 (2017): 82–116: 94-95. Chow writes that by adapting the international maritime law regarding state's contiguous zones, along with Article 17 of the United Nations Convention on the Law of the Sea (UNCLOS),

III. Warning Zones: This paper contends that a third type of zone would overcome the shortcomings of KOZs and SDZs. Although KOZs and SDZs have their merits, they suffer from two major drawbacks: first, our analysis finds that both violate international law; second, both are highly vulnerable to certain well-reasoned policy objections. Warning zones, if framed, designed, and implemented properly, would resolve these issues and objections. Warning zones, much like KOZs, would demarcate areas around certain vulnerable but strategically vital assets, and much like SDZs would emphasize the preemptive use of force in self-defense. However, unlike either KOZs or SDZs, the declaration of the zone itself is *not* understood—either implicitly or explicitly—to grant any right to the declaring state that it does not already possess. Instead, much like certain similar zones in the maritime domain, warning zones in space serve an information-gathering function: "trespass" *per se* is not restricted, but can, upon meeting certain thresholds, provide increasingly certain evidence of hostile intent which would justify preemptive use of force in self-defense. Warning zones, would, therefore, provide an important—indeed, essential—mechanism for clarifying intent, reducing the propensity for miscalculation by either side, improving signaling by both parties, and enhancing stability in crises.

Legality of Zones

Within this taxonomy, the critical distinction between zone proposals is their legality under international law. After the launch of Sputnik in 1957, the United Nations established an Ad Hoc Committee in 1959 to consider establishing legal rules to prevent state "interference among space vehicles."[325] Though the Committee sought to find analogous principles in maritime and airspace law to apply to outer space, the Committee acknowledged that outer space was distinguished "by many specific factual conditions," most important of which is that outer space is, "on conditions of equality, freely avail-

---

states would not be violation of the 'non-appropriation' and 'free access' principles of the OST.

325. Collins and Williams, "Towards Traffic Control Systems," 161–70.

able for exploration and use by all in accordance with existing or future international law or agreements."[326] It was these conditions, that no area of space was unavailable for free exploration or use by all states, that crystallized into the OST a decade later in 1967.

The OST has since set forth two key criteria that guide the legality of different zone proposals. First, Article I of the OST restricts the ability of a state to infringe on another state's "free access to all areas of celestial bodies" and mandates equality between states "without discrimination of any kind" (henceforth referred to as the 'free access' principle).[327] Second, Article II of the OST prohibits any state appropriation over outer space not only "by claim of sovereignty," but also "by means of use or occupation, or by any other means" (henceforth referred to as the 'non-appropriation' principle).[328]

---

326. Glenn Reynolds and Robert Merges, "Jessup & Taubenfeld, The United Nations *Ad Hoc Committee on the Peaceful Uses of Outer Space*, 53 American Journal of International Law 877 (1959)," in Outer Space: Problems Of Law And Policy (Routledge, 2019).
327. *Treaty on Principles Governing the Activities of States in the Exploration and Use of Outer Space, Including the Moon and Other Celestial Bodies* (OST), Jan. 27, 1967, 18 U.S.T. 2410, T.I.A.S. 6347, 610 U.N.T.S. 205 (effective Oct. 10, 1967).
328. *Treaty on Principles Governing the Activities of States in the Exploration and Use of Outer Space, Including the Moon and Other Celestial Bodies.*

Articles I and II, Text of the Outer Space Treaty

Article I	Outer space, including the moon and other celestial bodies, shall be free for exploration and use by all States without discrimination of any kind, on a basis of equality and in accordance with international law, and there shall be free access to all areas of celestial bodies.[329]
Article II	Outer space, including the moon and other celestial bodies, is not subject to national appropriation by claim of sovereignty, by means of use or occupation, or by any other means. [330]

Notably, there is firm international consensus on these principles. As of 2019, 109 countries are party to the OST, 89 of which are signatory states.[331] Strong evidence suggests that the 'non-appropriation' principle also constitutes customary international law. The origins of the document during the Cold War indicate the treaty was drafted as a general practice to prevent militarization of space.[332] Furthermore, the *travaux préparatoires* explicitly write that "[c]elestial bodies should not be subject to any claim of sovereignty."[333] Additionally, *A. D. Pershing* makes a strong case that the non-appropriation of space is *opinio juris* and broadly supported by states' domestic

329. Ibid.
330. Ibid.
331. Ibid.
332. Abigail D Pershing, "Interpreting the OST's Non-Appropriation Principle: Customary International Law from 1967 to Today," *The Yale Journal of International Law* 44, no. 1 (2019): 154-155.
333. Letter from Arthur Goldberg, Permanent Representative of the U.S., to the Chairman of the Comm. On the Peaceful Uses of Outer Space (June 16, 1966), http://www.unoosa.org/pdf/limited/c2/AC105_C2_L012E.pdf [hereinafter Letter from Arthur Goldberg] in Pershing, "Interpreting the OST's Non-Appropriation Principle: Customary International Law from 1967 to Today," *The Yale Journal of International Law* 44, no. 1 (2019): 155.

legislation.[334]

Consequently, the critical task for zone proponents is to overcome the 'free access' and 'non-appropriation' principles set forth by the OST and firmly held by the international community. As *T.J. Schwetje* wrote on the subject of zoning outer space in 1987:

> "The international community has generally not objected to the establishment of special zones if the zones under the circumstances were reasonable and did not unduly hamper or interfere with another nation's freedom to navigate the seas or supra-adjacent airspace. Important in the following discussion is the notion that the key to gaining international acceptance of such zones has been that they extend no claim of sovereignty. There is a clear distinction between sovereignty and the right to exercise a preventive, protective, or regulatory jurisdiction (Footnote omitted)."[335]

Consequently, the core question regarding the legality of zones centers on whether proposals constitute a *de facto* claim of sovereignty.

Unlike international maritime law, under which the United Nations Convention on the Law of the Sea clearly establishes the areas of water over which states exercise sovereignty, the limits of sovereignty over airspace are murky as no international agreement establishes an altitude limit on what constitutes "airspace."[336] State sovereignty over airspace dates back to the Roman Empire, but technological development over subsequent millennia, such as the advent of hot-air balloons and early airplanes, progressively increased the altitude over which states argued they could exercise effective control and claim sovereign right.[337] With the launch of Sputnik by the Soviet Union in 1957, the emergence of manmade satellites prompted sig-

334. Pershing, "Interpreting the OST's Non-Appropriation Principle," 157-162.
335. Schwetje, "Protecting Space Assets," 134.
336. Dean N Reinhardt, "The Vertical Limit of State Sovereignty," *Journal of Air Law and Commerce*. 72, no. I (2007): 65-137.
337. John C. Cooper, "High Altitude Flight and National Sovereignty," The International Law Quarterly 4, no. 3 (1951): 411–18: 412.

nificant legal questions about the bounds of airspace over which a state could exercise sovereignty.[338]

Out of concern that a boundary between airspace and outer space might constrain their military activities, neither the United States nor the Soviet Union pressed for official demarcation after the launch of Sputnik.[339] Prior to Sputnik, the Soviet Union held that no vertical limit on its airspace existed, but reversed this position to affirm the legality of Sputnik, arguing that no state could claim sovereignty over outer space.[340] The launch of Sputnik also incited internal debate within the United States about the vertical limit of United States airspace.[341] Essentially enshrining the principle of overflight into international law, the United States elected not to object to the Soviet position that Sputnik did not penetrate United States airspace.[342] From the perspective of the United States Air Force, this position allowed the United States to freely orbit the Soviet Union in its own space activities.[343] Furthermore, the lack of a clear delineation between airspace and outer space prevented "the direct assimilation of any pattern of conduct in space to the existing legal principle of airspace sovereignty, and thereby enables nations to approach space problems unfettered by prior commitments."[344]

---

338. Albert I Moon Jr, "A Look at Airspace Sovereignty," *Journal of Air Law and Commerce* 29, no. 4 (1963): 328–45.
339. M.V. Smith, "Ten Propositions Regarding Spacepower," The Fairchild Papers (Maxwell Air Force Base, Alabama: Air University, Air University Press, 2002): 4.
340. Reinhardt, "The Vertical Limit of State Sovereignty," 82-83.
341. Ibid., 85.
342. George M. Moore, Vic Budura, and Joan Johnson-Freese, Joint Space Doctrine: Catapulting into the Future, Joint Force Quarterly (Institute for National Strategic Studies, National Defense University, 1994): 74; Reinhardt, "The Vertical Limit of State Sovereignty," 82-88; The Harvard Law Review Association, "National Sovereignty of Outer Space," *Harvard Law Review* 74, no. 6 (April 1961): 1154–75: 1167.
343. Delbert R. Terrill Jr., "The Air Force Role in Developing International Outer Space Law" (Air University, Air University Press, 1999), https://www.airuniversity.af.edu/Portals/10/AUPress/Books/B_0069_TERRILL_OUTER_SPACE_LAW.pdf.
344. The Harvard Law Review Association, "National Sovereignty of Outer

The physics of Earth's atmosphere and orbit shape the practical exercise of sovereignty. Since the Earth's atmosphere gradually thins with altitude, there is no identifiable cut-point between airspace and outer space.[345] A number of arbitrary cut-points[346] have been pro-

---

Space," *Harvard Law Review* 74, no. 6 (April 1961): 1154–75: 1169.
345. Nadia Drake, "What Is the Kármán Line, and Where Is the Edge of Space?," National Geographic, 2018, https://www.nationalgeographic.com/science/2018/12/where-is-the-edge-of-space-and-what-is-the-karman-line/.
346. Stanley B. Rosenfield, "Where Air Space Ends and Outer Space Begins," Journal of Space Law 7, no. 2 (Spring & Fall 1979): 137-148. p.139-140:

> "These categories include:
>
> 1. The line based upon the concept of atmosphere, which has already been noted.
>
> 2. A line based on the division of the atmosphere into four layers, the troposphere, the stratosphere, the mesosphere and the ionosphere. Each division has its own scientific characteristics and each has given rise to a variety of different proposals, ranging from 31 miles above the earth to as high as 500 miles.
>
> 3. The ICAO theory previously: noted) based on the maximum altitude of aircraft flight.
>
> 4. The von Kifrman line, by which the line would be established at the point where aerodynamic lift yields to centrifugal force, which would put the line at approximately 275,000 feet.
>
> 5. The line based on the lowest perigee of an orbiting satellite. This is based on the fact that when the earth's atmosphere is too dense, an artificial satellite cannot remain in orbit.19
>
> 6. The line based on the point where the gravitational pull of the earth ceases. This theory arises from the basic assumption that for state security, the extent of sovereignty should extend beyond the point from which any object may be dropped.
>
> 7. The line between sovereign air space and outer space should be drawn at the limit of the underlying state's capacity to effectively apply its authority. This would provide a different limit on sovereign air space for different states.
>
> 8. The zonal theory under which air space would be divided

posed by both national and international organizations, but the lack of a physical boundary has spoiled attempts at concrete demarcation (see footnote).[347]

The distinction between airspace and outer space bears important consequence for both the 'free access' and 'non-appropriation' principles. Satellites in low-Earth orbit and celestial objects beyond Earth's orbit constantly orbit through areas of space belonging to other states and are unsuited to the vertical application of sovereignty.[348] However, a satellite in geosynchronous orbit, where it "appears to 'hover' over one spot on the Equator,"[349] could hypothetically "hover" within a state's borders on Earth. Eight equatorial countries in the 1976 Bogota Declaration claimed sovereignty over geosynchronous orbit: because everything within Earth's gravitational field along the equator constituted airspace directly above their geographic territories.[350]

---

into sovereign air space, based upon either the height at which aircraft can operate or some other line of demarcation; a contiguous zone, through which all non-military flights would have a right of free passage; and all that above contiguous space which would be free space.

9. Drawing the line between air space and outer space by a combination of one or more of the previously mentioned proposals.

347. See 3 and 9 for further discussion. For example, an initial 1978 proposal by the Soviet Union argued that the lowest possible perigee of a manmade satellite constituted outer space. However, the aforementioned political considerations led the United States and Soviet Union to operate on the premise that any satellite in orbit is in space and refused to offer an official boundary. Although this position did not clarify the boundary between airspace and outer space, it did clarify that satellites are in outer space and not airspace. For more, see Reinhardt, "The Vertical Limit of State Sovereignty," 116.
348. C.W. Jenks, International Law and Activities in Space, 5 Int'l & Comp. L.Q. 99, 103-104 (1956); Moon Jr, "A Look at Airspace Sovereignty," 335-337.
349. Mindi Capp/MSFC, "NASA - Geosynchronous Satellites," Feature Articles, Image Feature (Brian Dunbar), accessed May 4, 2020, https://www.nasa.gov/audience/forstudents/5-8/features/geo_feature_5_8.html.
350. Nayebi, Nima. 2011. "The Geosynchronous Orbit and the Outer Limits of Westphalian Sovereignty." *Hastings Science & Technology Law Journal* 3(2):

However, the sovereign declaration by these states was resoundingly rejected by the international community. Detractors argued that longstanding state practice during which the ITU delegated orbital slots to spacefaring nations was enshrined as custom.[351] Although the Bogota Declaration signatories asserted that geostationary orbit was "a physical fact linked to the reality of our planet because its existence depends exclusively on its relation to gravitational phenomena generated by the earth [sic]," these claims were also rejected.[352] Experts disagreed with the claim of exclusive dependence, stating that "geostationary orbit is dependent on several factors such as: the launch and station-keeping propulsion, the attraction of the earth, the moon and the sun, and the solar radiation pressure."[353]

Lastly, although the Bogota Declaration signatories argued that the 1967 OST did not define outer space (which, in their view, meant that one could not logically conclude geosynchronous orbit was in space),[354] it was already well established that satellites in outer space are not located in airspace. Not only was state practice by the United States and Soviet Union a testament to this fact, but the international community generally accepted that "airspace activities cannot take place beyond an altitude of sixty kilometers," which "would suggest that the sovereignty of airspace issue, based on the international law of the 1944 Chicago Convention, terminates at a level of sixty kilometers beyond which altitude, the Convention does not apply."[355]

471–98.

351. Ibid., 489.

352. *Declaration of the First Meeting of Equatorial Countries*. 1976. https://www.jaxa.jp/library/space_law/chapter_2/2-2-1 2_e.html.

353. Andrej Gorbiel, "The Legal Status of Geostationary Orbit: Some Remarks," Journal of Space Law 6, no. 2 (1978): 171–78: 176.

354. Michael J. Finch, "Limited Space: Allocating the Geostationary Orbit," Northwestern Journal of International Law & Business 7, no. 4 Fall (1986 1985): 788–802.

355. Ferdinand Onwe Agama, "Effects of the Bogota Declaration on the Legal Status of Geostation Orbit in International Space Law," Nnamdi Azikiwe University Journal of International Law and Jurisprudence 8, no. 1 (2017): 24–34; Furthermore, it has been argued that states lack effective control over such high altitudes which consequently states sovereignty over orbit, though this reasoning is unclear because states can take measures against craft penetrating their

Zones and Self-Defense

Whether on the basis of unlawful penetration or hostile approach, the unifying element among all zone proposals is the right to invoke self-defense. Consequently, it must be first established that the function of any type of zone, to defend against co-orbital ASATs, is legal under international law.

Although there exists a right to self-defense in outer space (see footnote),[356] this does not necessarily imply the permissiveness of any zone, whether they be KOZs, SDZs, or warning zones. These zones, however, must comport with requirements of necessity and proportionality. While proportionality in self-defense is almost by

---

airspace beyond targeting the craft itself (The Harvard Law Review Association, "National Sovereignty of Outer Space," *Harvard Law Review* 74, no. 6 (April 1961): 1154–75). In either case, the international community decided that the signatories of the Bogota Declaration had no reasonable claim of sovereignty over geosynchronous orbit.

356. Spanning only seventeen articles, the OST provides little more than a broad framework to govern space activities. However, in lieu of explicit provisions regulating specific activities, Article III states unequivocally that activities in outer space must be "in accordance with international law, including the Charter of the United Nations..." Accordingly, the explicit requirement that activities in space comport with international law, presupposes that outer space is a domain to which international law can be applied, as opposed to a lawless, Wild West completely liberated from international legal restraints. Such analysis is reflected in both state doctrine and practice. Recent research seeking to clarify existing international law at the intersection of outer space and military operations has been undertaken for the development of the *Woomera Manual Project (see Germany v. Denmark and the Netherlands* (also known as *The North Sea Continental Shelf* cases). 1969. International Court of Justice). The manual argues that Article III of the OST, which has been ratified and accepted by all signatories, explicitly applies international law and the United Nations Charter to outer space,. Furthermore, the International Law Commission (ILC) has concluded that provisions of international law that govern times of war, such as the Law of Armed Conflict (LOAC) is *lex specialis* during any period of armed conflict, regardless of the domain in which it occurs (Dale Stephens, summarizing the existing work of the Woomera Manual Project, "Military Space Operations and International Law," *The Woomera Manual Project – Part 1.* March 2, 2020) As a result, the legal basis for preemptive self-defense in terrestrial domains also demonstrates its legality in outer space.

definition context-dependent, contemporary jurisprudence identifies one or both of two tests: force at the level of equivalence of the initial attack, and force required for cessation of the attack.[357] Warning zones clear both thresholds. As such, a nation choosing to disable or destroy a "stalking" enemy satellite has acted proportionally, provided they limit, to the extent possible, the response solely to the satellite in question, and avoid "cross-domain" escalation. If bodyguard satellite technologies prove effective and feasible, a proportional response would likely be to disable solely the satellites that have violated a warning zone. If these technologies, however, do not offer an effective defense against co-orbital ASATs, the United States will need to consider other options, such as holding an adversary's critical satellites at risk and threatening retaliation.

The first potential threshold for proportionality is the equivalence, or symmetry requirement: the defender, when acting to dispel the aggressor, must roughly approximate the force level of the attack. Such a restriction is a necessary component of *jus ad bello*; otherwise, in the view of Professor *Yoram Dinstein*, the most minor of skirmishes could result in an asymmetric response intended to fell the entire forces of the attacker themselves.[358] International custom supports this premise. In the case of the *Caroline* incident,[359] the

357. Etezazian, S., "The nature of the self- defense proportionality requirement," *Journal on the Use of Force and International Law*, 3(2), 2016. 260–289.
358. Dinstein, Yoram, "War, Aggression and Self-Defence,"*Cambridge University Press*, 5th edn 2011. 262–63.
359. In December of 1837, forces under the authority of Great Britain set ablaze the *Caroline*, an American vessel docked in U.S. shores at Schlosser's Landing, destroying it. The *Caroline* housed a group of Canadian separatists seeking to foment political unrest. Claiming that the vessel posed a threat to Canadian (and, subsequently, British) interests, Great Britain argued that there existed a right to act against that which could pose a future threat, prior to the actual occurrence of such threat. After the incident, a statement was issued by United States Secretary of State Daniel Webster that, to justify the invocation of self-defense, a state must demonstrate "a necessity of self-defense, instant, overwhelming, leaving no choice of means, and no moment deliberation" (see Historical record of correspondence between Secretary Webster and Lord Ashburton provided by the Yale Law School Lillian Goldman Law Library). Over the past several centuries, this has crystallized into the principle of anticipatory self-defense, which relies on

British thought a proportional response to the perceived threat posed by forces on the *Caroline* was elimination of the ship, a roughly symmetrical response; there would have been no entertaining of a land-invasion in response.

With ample warning and *ex ante* clarification, KOZs, SDZs, and warning zones are proportional. Zones set a series of specific red-lines that, if crossed, provide information to the defender that they should expect an attack. If the United States informs adversaries that it expects there to be, for example, 50 kilometers separating United States' satellites from foreign spacecraft, and that encroaching would be an act of war, adversaries knowingly crossing into those zones communicates to the defender that a threshold has been crossed. For the defending state to presume intentional aggression and dispel in an anticipatory matter is eminently reasonable. Such warning epitomizes a tit-for-tat response: in response to an enemy satellite threatening to destroy a nation's satellite, the defending state destroys the offending satellite. In such a case, the response would likely be limited solely to the tailgating satellite in question. Compared to the elegant simplicity of a response that treats a threat against a satellite as warranting destruction of the attacker's satellite, cross-domain responses, however, (which range from economic coercion such as sanctions to military responses such as ground strikes) would be inappropriate and not analogous.

The second potential threshold for proportionate self-defense concerns whether it's aimed merely at dispelling the attacker, the "Cessation Approach." Such a right of self-defense solely provides the grounds for an anticipatory response to the attacker. Upon disabling the threat, the prosecution of the war effort ends, as the objective of self-defense has been accomplished.[360] This approach also finds support in existing custom. In the *Nicaragua*[361] dissent by Judge

---

two conditions: necessity (the lack of other, countervailing options), and proportionality (not exceeding the threshold needed to respond to the existing threat).

360. Antonio Cassese, "Terrorism Is Also Disrupting some Crucial Legal Categories of International Law," *European Journal of International Law*, Vol 12. 993. 2001.

361. *The Republic of Nicaragua v. The United States* (1986), decided by the ICJ,

Schwebel, it's noted that "It would be mistaken ... to think that there must be proportionality between the conduct constituting the armed attack and the opposing conduct. The action needed to halt and repulse the attack may well have to assume dimensions disproportionate to those of the attack suffered....[362]" As with the symmetrical threshold, it is clear that warning zones do not run afoul of the Cessation Approach. The anticipatory action is aimed solely at disabling the attacking satellite, with no corresponding military campaign taking place either in outer space or terrestrially.

More broadly, the previous cases examined have a common throughline: while proportional responses will depend on the situation, they will always be defined by a use of force narrowly tailored to a potential threat. In the case of the *Caroline*, acting against the ship was justified; invading the United States would not be. Similarly, actions in space to respond to the aggressive, stalking satellite are narrowly tailored, only using a level of force that is both equivalent to the initial threat and limited to dispelling the attacker. Additionally, part of

---

touched on the precise issue. Holding the United States at fault for supporting the Nicaraguan Contras against the Sandinistas government, the court found a violation of sovereignty via intervention in the internal affairs of another state through an arms trade (see *The Republic of Nicaragua v. The United States of America* (1986). International Court of Justice). Regardless of the specific point at issue, in evaluating the American claim that arms sales to the Contras were an act of self-defense, the court drew a useful extrapolation from the case to apply to self-defense more broadly. In his dissent, United States Judge *Stephen Schwebel* argued that no caveat exists in Article 51 which evidences that the right to self-defense may be invoked "if, and only if, an armed attack occurs." Rather, the phrasing of an inherent right makes it applicable, regardless of circumstance, instead of merely being prescribed to an *ex post* response (See Anthony Clark Arend, "International Law and the Preemptive Use of Military Force," *The Washington Quarterly*. Volume 26, Spring 2003. pp. 89-103). Additionally, other sections of the UN Charter lend credence to such interpretation. Article 2 prohibits the "threat or use of force" (Article 51 of the Charter of the United Nations. June 6, 1945) against member states; notable is the prohibition not just on the aggressive action itself, but a prohibition on events that may portend future aggression.

362. *The Republic of Nicaragua v. The United States of America* (1986). International Court of Justice.

the inquiry in *Nicaragua* may offer a useful insight: the threshold of reasonableness. Judge *Schwebel's* opinion alluded to a "reasonableness" inquiry as part of self-defense, writing that there exists "a right to take such actions as are *reasonably* necessary to end the attack promptly and protect the threatened values" [Emphasis is ours].[363] While there may be no cut-and-dry method, if a threat poses a grave danger to the national interests of a state, the state's response should be accorded some leeway provided that it appears justified in light of the threat posed. A threat in space to multiple critical satellites, with a symmetrical response, surely clears such a threshold. As a result, any zone proposal must establish clear boundaries such that, if penetrated by another state, confirm the use of force in self-defense.

Keep-Out Zones

Proposals to establish KOZs, which restrict close approaches by declaring zones around sensitive space assets in which unauthorized entry into or passage through which may be met with military force, are in clear violation of both the 'free access' and 'non-appropriation' principles. In its initial proposal, the OTA acknowledged that the 'free access' and 'non-appropriation' principles presented the primary barrier to KOZs, because they would constitute the appropriation of outer space to defend against anti-satellite weapons. The OTA's counterargument to this criticism was that the allocation of the geosynchronous orbital slots by the ITU "already incorporates a variation of the 'keep-out zone' principle," because satellites must be placed several angular degrees apart to avoid interference, and, that this "precludes the placement of other satellites near its position in the orbital arc."[364] However, there are a variety reasons as to why the ITU does not provide legal precedent for KOZs.

By way of background, the allocation of orbital slots by the ITU

---

363. *The Republic of Nicaragua v. The United States of America* (1986). International Court of Justice.
364. "Anti-Satellite Weapons, Countermeasures, and Arms Control" (Washington: U.S. Congress, Office of Technology Assessment, September 1985), 118.

originates from the designation of geostationary orbit along the equator as a "limited natural resource" in Article 33 of the International Telecommunication Convention (ITC) due to limited orbits available and concerns about radio frequency interference between satellites. As a result, the International Telecommunication Union assumed responsibility for the allocation of orbital slots to spacefaring nations to reduce radio frequency interference.[365]

It is firmly established that the allocation of orbital slots by the ITU does not constitute a violation of the 'free access' or 'non-appropriation' principles of the OST. First, although orbital slots do, in effect, create a zone in which no state (other than the state to which the slot is allocated) can place geosynchronous satellite, the ITU does not limit a state's the 'free access' to an orbital slot. Rather, all states are afforded freedom of access on a "first come, first served" basis and are not granted any proprietary rights over the orbital slot.[366] Although states are unable to place satellites in orbital slots not specifically allocated to them by the ITU, there is absolutely nothing that permits a state from preventing infringement through the use of force. As a result, there is no practical equivalence between orbital slots and KOZs as proposed by the OTA.

Second is the nature of the actor involved. Unequivocally, Article II prohibits the 'national appropriation' of outer space, such as by nations including the United States and Russia. In the case of the OTA's proposal, it suggests that the United States unilaterally declare KOZs surrounding United States space assets or negotiate these zones through a treaty. [367] Consequently, to analogize the United States to the ITU in this context is incorrect. As an international organization, the ITU is meaningfully distinct from a na-

365. Collins and Williams, "Towards Traffic Control Systems," 161–70.
366. Elina Morozova and Yaroslav Vasyanin. 2019. "International Space Law and Satellite Telecommunications." *Oxford Research Encyclopedia of Planetary Science*. https://oxfordre.com/planetaryscience/view/10.1093/acrefore/9780190647926.001.0001/acrefore-9780190647926-e-75 (May 7, 2020).
367. "Anti-Satellite Weapons, Countermeasures, and Arms Control" (Washington: U.S. Congress, Office of Technology Assessment, September 1985), 8.

tion.[368] Although some have argued that the "process of allocating orbital space and excluding nations from particular slots seems to be an obvious case of appropriation," the ITU has broad international character and preserves "access for all space actors, current or future."[369] This fact explains why the ITU has never been abolished or restricted from its activities allocating orbital slots.[370]

Next is the question of what constitutes 'appropriation.' In Stephen Gorove's article, *Major Legal Issues Arising from the Use of the Geostationary Orbit*, Gorove defines "appropriation" as "the taking of property for one's own or exclusive use with a sense of permanence."[371] Gorove further writes that—

> "While a state may certainly exercise exclusive control over a traditional object, such as a ship, or an aircraft, or a part of airspace, it is not clear that a satellite in geostationary orbit would be able to maintain its exact position and occupy the same area over a period of time. Even if a position could be accurately maintained, and thus possibly constitute an "appropriation" within the meaning of article II, the satellite would have to be kept in that orbit with a "sense of permanence" and not on a temporary basis."[372]

This demonstrates two reasons as to why KOZs and orbital slots are

---

368. Dietrich Westphal, "International Telecommunication Union (ITU)," Oxford Public International Law, accessed May 8, 2020, http://opil.ouplaw.com/view/10.1093/law:epil/9780199231690/law-9780199231690-e514.
369. Timothy Justin Trapp, "Taking Up Space By Any Other Means: Coming To Terms With The Nonappropriation Article Of The OST," *Illinois Law Review* 2013 (2013): 1681–1714: 1709-1710.
370. Trapp, "Taking Up Space By Any Other Means: Coming To Terms With The Nonappropriation Article Of The OST," 1709-1710.
371. S. Gorove, Studies in Space Law: Its Challenges And Prospects 82 (1977) in Stephen Gorove, "Major Legal Issues Arising from the Use of the Geostationary Orbit," *Michigan Journal of International Law* 5, no. 1 (January 1, 1984): 3–12: 5.
372. Gorove, "Major Legal Issues Arising from the Use of the Geostationary Orbit," 5.

meaningfully distinct. First, all states are afforded freedom of access to an orbital slot on a "first come, first served" basis and are not granted any proprietary rights over the orbital slot— they are, both legally and in effect, impermanent.[373] Second, the categorization of geostationary orbit as a "limited natural resource" in Article 33 of the International Telecommunication Convention (ITC) shields orbital slot allocation from claims of 'appropriation' by the ITU.[374] Authority on this question can be found in the International Law Association's Report of the Fifty-Fourth Conference 427 (1970), along with the United Nations Committee on the Peaceful Uses of Outer Space, which have suggested that 'appropriation' does not apply to natural resources.[375]

Although KOZ proponents do not advocate that states claim sovereignty over the zones themselves, and therefore meet the "by claim of sovereignty" qualification in Article II, this does not elide Article II's limitation on appropriation by "means of use or occupation."[376] This is perhaps the most clear indication of KOZs illegality under the OST, because the basis for limiting another state's 'free access' to the zone itself is its 'use' by a state's satellite. For example, the basis for the OTA's argument about the legality of 'keep-out zones' rests on the notion that, based upon ITU regulations, "geosynchronous [orbits] must be space several degrees apart in order to avoid frequency interference. Therefore, such a satellite precludes the

373. Morozova and Vasyanin, "International Space Law and Satellite Telecommunications," 2019; Mark Holmes, "Hot Orbital Slots: Is There Anything Left?," Via Satellite, March 1, 2008, https://www.satellitetoday.com/uncategorized/2008/03/01/hot-orbital-slots-is-there-anything-left/; Lotta Viikari, *The Environmental Element in Space Law: Assessing the Present and Charting the Future* (Martinus Nijhoff Publishers, 2008).
374. Ibid., 6; Collins and Williams, "Towards Traffic Control Systems," 161-170.
375. INTERNATIONAL LAW ASSOCIATION, REPORT OF THE FIFTY-FoURTH CONFERENCE 427 (1970). This position has also been adopted by UNCOPUOS in reference to the natural resources of the moon and other celestial bodies. See UNCOPUOS, Doc. A/AC.105/C.2/WG.1/WP.3; Press release U.S.-U.N. 37 (73) in Goroves, "Major Legal Issues Arising from the Use of the Geostationary Orbit," 12.
376. Ibid.

placement of other satellites near its position in the orbital arc."[377] In other words, the usage of the orbital slot precludes its access and usage by another state. As we have already established, the ITU is exempt from Article II in the context of geosynchronous orbit. However, for a state to declare a 'keep-out' zone on the basis that it preserves national security, as *Blau* and *Gore* propose, this would certainly constitute national appropriation "by use" according to the text of Article II.[378]

Other proponents of KOZs note that precedent exists in the four-kilometer "nominal approach" zone and 200-meter protective "'keep-out' zone" around the International Space Station (ISS).[379] Unlike KOZs, however, the zones around the ISS do not permit any state to enforce the zone, prevent entry, or invoke the right to self-defense. Furthermore, states aboard the ISS have not claimed sovereignty over the area or sought to exclude other states from entering the area.[380] Instead, the zones around the ISS are "guidelines that are used to assess whether the threat of such a close pass is sufficient to warrant evasive action or other precautions to ensure the safety of the crew."[381] As a result, these zones, much like warning zones, can be interpreted as an information-gathering mechanism as opposed to an exclusive measure to allow the invocation of self-defense.

377. "Anti-Satellite Weapons, Countermeasures, and Arms Control" (Washington: U.S. Congress, Office of Technology Assessment, September 1985), 118.
378. Schwetje, "Protecting Space Assets," 133.
379. Rebecca Reesman and Andrew Rogers, "Getting in Your Space: Learning from Past Rendezvous and Proximity Operations" (The Center for Space Policy and Strategy, The Aerospace Corporation, May 2018), https://aerospace.org/sites/default/files/2018-05/GettingInYourSpace.pdf.
380. Newsome, "The Legality of Safety and Security Zones in Outer Space: A Look to Other Domains and Past Proposals" (Montreal, Quebec, McGill University, Institute of Air and Space Law, 2016), 66-69.
381. Mark Garcia, "Space Debris and Human Spacecraft," Text, NASA, 2013, http://www.nasa.gov/mission_pages/station/news/orbital_debris.html.

Self-Defense Zones

Unlike KOZs, SDZs bear resemblance to zones on Earth, such as *B.G. Chow's* self-defense zone and *Lt. Col. Butler*'s "Space Defense Identification Zone" strategies, yet previous proposals would similarly violate Articles I and II of the OST.[382] *T.A. Newsome* explains that Air-Defense Identification Zones (ADIZs), perhaps the most comparable terrestrial zone to SDZs, "represent the most direct challenge to Article II's prohibition on national appropriation."[383] Similar to SDZ proposals, ADIZs on Earth do not establish any sovereign right, can be established unilaterally, rely on voluntary compliance, permit overflight, and, most importantly, can permit the right to self-defense. However, much like states are prohibited from claiming sovereignty over outer space, the Convention on the High Seas prohibits states from claiming sovereignty over the high seas.[384] Could this state practice justify the establishment of SDZs?

To propose that ADIZs legally justify SDZs would be unreasonable. Although ADIZs themselves do not create a sovereign right over airspace, the legality of ADIZs extend from preexisting state sovereignty over a state's national airspace and territorial seas.[385] Importantly, ADIZs cannot legally extend beyond areas over which states do not already claim sovereign right. To this effect, the Convention on the Territorial Sea and the Contiguous Zone and the Convention on the High Seas prohibit states from claiming sovereignty over the high seas.[386] Importantly, ADIZs are grounded in customary international law, so there is no convention that permits the legality of ADIZs over the high seas for a special purpose, such as national

---

382. Ibid., 133.
383. Newsome, "The Legality of Safety and Security Zones in Outer Space," 66-69.
384. Elizabeth Cuadra, "Air Defense Identification Zones: Creeping Jurisdiction in the Airspace," *Virginia Journal of International Law* 18, no. 3 (1978 1977): 485–512: 499.
385. "Air Defence Identification Zones," Oxford Public International Law, accessed May 7, 2020, http://opil.ouplaw.com/view/10.1093/law:epil/9780199231690/law-9780199231690-e237.
386. Cuadra, "Air Defense Identification Zones," 499.

security.[387]

International consensus on the inappropriateness of China's ADIZ in the East China Sea (ECS) demonstrates the limited applicability of ADIZs to SDZs. In 2013, China announced the establishment of the "East China Sea Air Defense Identification Zone," which covered large swaths of the ECS, including disputed areas between China and Japan.[388] The United States, Japan, South Korea, and Taiwan criticized China's ADIZ as escalatory, and Japan argued that the zone was a violation of international law. [389] Legal experts have rejected China's ADIZ in the ECS. For example, *R.G. Almond* explains that:

> "ADIZs can only be legally applied in relation to preventing the unauthorized entry of aircraft into the national airspace. ADIZs cannot be used to control foreign aircraft not intending to enter the national airspace. States only enjoy exclusive sovereignty over the airspace above their territory, a right which ends at the 12 nautical mile border of the territorial sea. Beyond this territorial belt, all states enjoy the high seas freedoms, including freedom of overflight, a customary principle memorialized in UNCLOS."[390]

As a result, China's ADIZ does not permit them to disallow overflight by other states and invoke the right to self-defense if such overflight occurs. Furthermore, the Convention on International Civil Aviation, the treaty whose principles guide the implementation of ADIZs, was amended in 1983 to prohibit the use of weapons against civilian aircraft.[391]

---

387. Cuadra, "Air Defense Identification Zones," 500.
388. Ian E Rinehart and Bart Elias, "China's Air Defense Identification Zone (ADIZ)," Congressional Research Service Report, 2015, 45: 6-7.
389. Rinehart and Elias, "China's Air Defense Identification Zone (ADIZ)," 18-22.
390. Roncevert Ganan Almond, "South China Sea: The Case Against an ADIZ," 2016, https://thediplomat.com/2016/09/south-china-sea-the-case-against-an-adiz/.
391. Almond, "South China Sea: The Case Against an ADIZ," 2016.

The conclusion of the Permanent Court of Justice in the *Lotus* case further complicates the ability for states to claim some competence or special jurisdiction in SDZs. As described by *Z. Papp*—

> "...international law does not prohibit a state from exercising (prescriptive) jurisdiction in its own territory in respect of any case which relates to acts that have taken place abroad. According to the findings of the Court, the states' corresponding discretion is solely limited by prohibitive rules of international law. As regards exercising enforcement powers, it cannot be exercised by a state outside its territory except by virtue of a permissive rule derived from international custom or from a convention."[392]

Consequently, for a state to lawfully create an SDZ, it would need to establish that the zone itself was not prohibited by international law, and that the zone was enforceable outside the state's territory due to some permissive rule.

However, Article II of the OST prohibits a state from establishing these conclusions. This brings us to the third qualification to the 'non-appropriation' principle in Article II that national appropriation is prohibited "by any other means."[393] Despite the fact that states only exercise special jurisdiction in ADIZs over the high seas, *Cuadra* writes the "competence claimed here, however, is one of limited sovereignty in airspace over waters beyond the territorial seas, for the specific purpose of 'national security.'"[394] Demonstrating that these principles would constitute national appropriation according to Article II of the OST, the exercise of "limited sovereignty" through an SDZ would certainly constitute national appropriation, because such a claim would be tied to a specific purpose or competence, such

392. Zoltan Papp, "Air Defense Identification Zone (ADIZ) in the Light of Public International Law," Pecs Journal of International and European Law 2015, no. 2 (2015): 28–54: 33-34.
393. "Treaty on Principles Governing the Activities of States in the Exploration and Use of Outer Space, including the Moon and Other Celestial Bodies". United Nations Office for Disarmament Affairs. Accessed May 8, 2020.
394. Cuadra, "Air Defense Identification Zones," 499.

as national security, that is explicitly prohibited under the "any other means" qualification in Article II of the OST. Furthermore, without some special purpose or claim of limited sovereignty, states could not invoke the right to self-defense. For example, *Blau* and *Gore* propose that the penetration of an established zone in space would constitute a threat, but *Schwetje* offers no evidence to suggest that this would constitute some permission allowing or circumventing the issue of appropriation.[395]

Customary international law which undergirds ADIZs would also not be applicable to orbit. As *Cuadra* describes—

> "...any legal basis for ADIZ's extending over the high seas must derive from some aspect of customary international law that is so fundamental a principle as to prevail over the will of the community of nations as expressed in these conventions."[396]

There is consensus that ADIZs are beginning to crystallize into customary international law based upon decades of state practice.[397] However, there is no comparable customary international legal basis for the appropriation of areas around a satellite based upon state practice.[398]

This presents three major obstacles to the application of ADIZ principles to outer space. Considering that ADIZs draw from customary international law, but are limited in scope by existing international conventions, there must be some analogous custom that would permit the establishment of SDZs. Our research has identified no such custom. Furthermore, the sweeping and explicit prohibition on national appropriation in Article II of the OST limits the ability of the state to take advantage of this lacuna. Second, considering that states have no sovereign claim over outer space, which is firmly

---

395. Schwetje, "Protecting Space Assets," 133.
396. Cuadra, "Air Defense Identification Zones," 500-501.
397. Ibid., 485; Papp, "Air Defense Identification Zone (ADIZ) in the Light of Public International Law," 47-48.
398. See previous section.

established by the OST and international rejection of the Bogota Convention, states cannot reasonably limit transversal of an SDZ because they are not preventing a spacecraft from entering their airspace. As explained by *Almond*, ADIZs "can only be legally applied in relation to preventing the unauthorized entry of aircraft into the national airspace."[399] Third, to overcome the aforementioned two arguments, states would need to claim some special competence over outer space, such as for the purpose of national security, which is expressly prohibited by Article II of the OST.

Warning Zones

Overcoming the aforementioned barriers to previous zone proposals, the application of the concept of maritime warning zones to outer space offers an intriguing solution to the dangers posed by co-orbital ASATs. In a maritime context, warning zones serve as "Defense bubbles" surrounding naval vessels.[400] As explained in the United States Commander's Handbook on the Law of Naval Operations—

> "As States endeavor to protect their interests in the maritime environment during peacetime, naval forces may be employed in geographic areas where various land, air, surface, and subsurface threats exist. Commanders are then faced with ascertaining the intent of persons and objects (e.g., small boats, "low slow flyers," jet skis, swimmers) proceeding toward their units. In many instance [sic] ascertaining their intent is very difficult, especially when operating in the littorals where air and surface traffic is heavy. Given an uncertain operating environment, commanders

399. Almond, "South China Sea: The Case Against an ADIZ," 2016.
400. Wolff Heintschel von Heinegg, "Current Legal Issues in Maritime Operations: Maritime Interception Operations in the Global War on Terrorism, Exclusion Zones, Hopsital Ships and Maritime Neutrality," *International Law Studies* 80, no. Issues in International Law and Military Operations (2006): 207–33: 213.

> may want to establish some type of assessment, threat, or warning zone around their units in an effort to help sort the common operational picture and ascertain the intent of inbound entities. This objective may be accomplished during peacetime while adhering to international law as long as the navigational rights of other ships, submarines, and aircraft are respected."[401]

Much like Article I of the OST, international maritime law does not recognize any situation during which freedom of navigation on the high seas can be limited. However, warning zones are neither operational nor exclusionary, and instead "merely serve to protect the naval vessels from attack or from illegal activities..."[402] Although these zones are historically established during wartime or national emergency, it is generally accepted that these zones can be established during peacetime under international law to protect naval vessels.[403] As *W. H. von Heinegg* writes, "the extent of these zones and the measures taken cannot be determined *in abstracto*. Rather, it will depend on the circumstances of each single case, especially on a known threat and on the location of the ships concerned, and whether the extent of the warning area may be reasonable or excessive."[404]

In outer space, the establishment of these zones on a case-by-case basis under threatening circumstances is particularly important. There should be two central considerations: the distance of attacking satellite, and the number of defending satellites being tailgated. The concurrent fulfillment of *both* of these conditions should suffice to allow an armed response. While this paper does not propose a specific number, leaving that to be determined through diplomatic negotiations, there should be an agreed-upon distance, or zone of separation between satellites. If a foreign satellite crosses within

---

401. "Safeguarding U.S. National Interests in the Maritime Environment," in The Commander's Handbook on the Law of Naval Operations, August 2017 (Navy Warfare Development Command, Norfolk, VA: Department of the Navy Office of the Chief of Naval Operations, 2017), 4-9.
402. von Heinegg, "Current Legal Issues in Maritime Operations," 213-214.
403. Ibid., 214.
404. Ibid., 214.

that zone, that clears the "distance" condition. The second condition is triggered by multiple satellites being followed within the zone of separation. Demonstrating warning zones' compliance with Article II of the OST, warning zones can be established without any operational or exclusionary purpose beyond the protection of the vessel itself.[405] As a result, the establishment of warning zones does not confer upon the state any additional right, or employ any special purpose jurisdiction, under international law.[406] In other words, a warning zone does not permit a state to violate freedom of navigation and forcibly remove any vessel from the zone. However, the zone itself does provide a means for states to ascertain or clarify the intent of nearby traffic and whether the invocation of self-defense may be necessary.[407]

Ascertaining intent is at the heart of addressing the risks posed by co-orbital ASATs. Chow writes that prohibiting passage through a zone around a satellite is impossible because of the difficulty distinguishing between co-orbital ASATs and civilian satellites, and the necessity of these forms of passage for regular satellite operations.[408] In a maritime context, warning zones allow ships to navigate a similar dilemma. In the global War on Terror, it can be difficult for navies to distinguish between threatening and peaceful civilian seacraft, so warning zones have been applied by navies to ascertain the intent of seacraft as they pass through a warning zone.[409] By establishing a "defense bubble" around the naval vessels, warning zones allow commanders to ascertain the intent of incoming seacraft, whether they be warships, small boats, or jet skis.[410] Furthermore, by direct-

---

405. Jon M. Van Dyke, "Military Exclusion and Warning Zones on the High Seas," *Marine Policy* 15, no. 3 (May 1, 1991): 147–69, https://doi.org/10.1016/0308-597X(91)90059-K; von Heinegg, "Current Legal Issues in Maritime Operations," 213-214.
406. Newsome, "The Legality of Safety and Security Zones in Outer Space," 33-34.
407. Ibid., 33-34.
408. Chow, "Stalkers in Space," 95-99.
409. von Heinegg, "Current Legal Issues in Maritime Operations," 215.
410. "Safeguarding U.S. National Interests in the Maritime Environment," in The Commander's Handbook on the Law of Naval Operations, 4-9.

ing all maritime traffic around the zone and requiring that all traffic passing through the zone communicate their intentions, commanders are able to clarify any threats and determine the need to invoke self-defense.[411] Upon notice, state or civilian seacraft might avoid entering the zone itself or choose to do so while communicating their intent to the United States Navy.[412]

The unilateral establishment of warning zones around United States satellites presents a potential solution to the threat of co-orbital ASATs without violating Articles I and II of the OST. First, the establishment of warning zones would not limit 'free access' to the area of the zone as specified by Article I. According to the general practice of warning zones, the establishment of the zone itself does not limit another state from entering the area. Furthermore, the penetration of the zone by another state does not license the use of force by the state enforcing the zone. As explained by the Commander's Handbook—

> "Specifically, when operating in international waters, commanders may assert notice (via notices to airmen and notices to mariners) that within a certain geographic area for a certain period of time dangerous military activities will be taking place. Commanders may request that entities traversing the area communicate with them and state their intentions. Moreover, such notice may include reference to the fact that if ships and aircraft traversing the area are deemed to represent an imminent threat to US naval forces, they may be subject to proportionate measures in self-defense. Ships and aircraft are not required to remain outside such zones and force may not be used against such entities merely because they entered the zone. Commanders may use force against such entities only to defend against a hostile act or demonstrated hostile intent, including interference with declared military activities."[413]

411. Ibid.
412. Ibid.
413. Ibid.

Warning zones would also avoid the 'national appropriation' principle under Article II. Referring back to the initial three qualifications of the treaty, the first of these qualifications is that national appropriation is prohibited "by claim of sovereignty."[414] As our previous analysis suggests, warning zones would not constitute a claim of sovereignty because they do not grant any sovereign right over the area of outer space. As a result, we direct our analysis towards the secondary and tertiary qualifications of what constitutes national appropriation "by means of use or occupation," or "by any other means."[415]

Proponents of 'keep-out' zones have cited ITU precedent for the limitation of space traffic through areas of outer space, arguing that its allocation by the ITU and use by a satellite is a limiting factor.[416] However, our previous analysis suggests that the OST restricts this interpretation through its second qualification to the prohibition on national appropriation "by means of use."[417] Warning zones, however, circumvent this problem. Unlike maritime warning areas, which predicate the establishment of the zone on its use for military activities, warning zones are not established on an operational basis associated with their use.[418] Instead, the establishment of the warning zone is predicated on a potential threat posed to the naval vessel. Applied to an orbital context, a threatening configuration of co-orbital ASATs would permit the establishment of a warning zone on the basis "that if ships and aircraft traversing the area are deemed to represent an imminent threat to US [forces], they may be subject to proportionate measures in self-defense."[419]

---

414. *Treaty on Principles Governing the Activities of States in the Exploration and Use of Outer Space, Including the Moon and Other Celestial Bodies* (OST), Jan. 27, 1967, 18 U.S.T. 2410, T.I.A.S. 6347, 610 U.N.T.S. 205 (effective Oct. 10, 1967).
415. Ibid.
416. "Anti-Satellite Weapons, Countermeasures, and Arms Control" (Washington: U.S. Congress, Office of Technology Assessment, September 1985), 118.
417. "Anti-Satellite Weapons, Countermeasures, and Arms Control," 118.
418. von Heinegg, "Current Legal Issues in Maritime Operations," 213-214.
419. "Safeguarding U.S. National Interests in the Maritime Environment," in The Commander's Handbook on the Law of Naval Operations, 4-9.

The next qualification concerns appropriation of outer space "by any other means."[420] In this context, warning zones offer a comparatively better solution than SDZs. As we have previously explained, the basis for the establishment of SDZs would be on some permissive right or special purpose such as national security. However, *Cuadra's* analysis indicates that this basis entails a limited form of sovereignty,[421] which would run afoul of the prohibition on appropriation by "any other means," if not by a "claim of sovereignty."[422] Warning zones, on the other hand, circumvent this concern because no permissive right or special purpose is established, but the right to self-defense is predicated off threatening approaches by other states.[423] As a result, states "may use force against such entities only to defend against a hostile act or demonstrated hostile intent, including interference with declared military activities."[424]Some may object that anticipatory responses in space fail to clear the threshold for imminence required by the "necessity" prong of self-defense. While there exists no one-to-one relationship between "pre-positioning" a satellite near another and an attack, there nonetheless is little reason to loiter near multiple foreign satellites intentionally except to mount an attack when the situation calls for it. This, clearly, exists as a preparation for an armed attack and justifies anticipatory responses. In geostationary orbit (GEO), where critical satellites such as nuclear early warning systems are located, all objects are stationary relative to other objects, with a fixed orbital slot. Remaining in place in GEO is comparatively easier than in other orbits, such as low-earth orbit (LEO). As such, ascertaining intent behind deliberate maneuvering in GEO is easier than in other orbits; a foreign satellite suddenly moving towards a nation's own satellite in GEO is very likely an intentional step. Now, consider the pre-positioning of foreign satellites right up against multiple satellites owned by

420. Ibid.
421. Cuadra, "Air Defense Identification Zones," 499.
422. *Treaty on Principles Governing the Activities of States in the Exploration and Use of Outer Space, Including the Moon and Other Celestial Bodies.*
423. "Safeguarding U.S. National Interests in the Maritime Environment," in The Commander's Handbook on the Law of Naval Operations, 4-9.
424. Ibid.

a nation. GEO, unlike other orbits, is exceptionally distant, with a significant lag time between launches and objects reaching orbit. If, in peacetime, an enemy had pre-positioned dozens of their own satellites right up against our most critical satellites, the United States would have no choice but to watch helplessly at the onset of a conflict as they disabled satellite after satellite.

Regardless of the precise contours of one's view on proportionality, it is clear that warning zones are both symmetrical and limited solely to the cessation of the enemy's aggression against the defender's satellites. This makes clear that such zones are an appropriate, measured response undertaken for the purpose of anticipatory self-defense, without running afoul of *jus in bello.* The trailing or "stalking" of one's satellites serves as a grave threat to the territorial integrity of a nation, clearing the thresholds of necessity and imminence for a *jus ad bellum* determination that anticipatory action is needed; such is consistent with decades of custom, backed by the imprimatur of Article 51, and is as applicable in space as in any terrestrial domain.

Policy Rationale for Zones

<u>Defining Space Stalking</u>

The threat posed by co-orbital ASATs remains new and ill-defined by both policymakers and academics. Few scholars have attempted to define what qualifies as tailgating, yet the policy community is beginning to take the threat of co-orbital anti-satellite weapons seriously again. The lack of agreement demonstrates that states must take proactive measures to discuss what behaviors qualify as tailgating and which do not. Russia and China have, on multiple occasions, conducted close approaches but have refused to acknowledge that these behaviors could be considered escalatory or provocative. In one 2015 incident, Russia's Luch satellite parked itself in-between two commercial satellites operated by Intelsat and refused to respond to multiple requests by the company for Russia to clarify its behavior.[425] More recently, the Russian Cosmos 2542 and 2543 satellites have closely mirrored the orbit of a National Reconnaissance Office spy satellite without clarifying their behavior, demonstrating the potential for Russia to weaponize advancing RPO technologies.[426] International dialogue to recognize tailgating as a hostile and provocative act would lay the groundwork for future negotiations to enhance the stability of the space domain.

Dr. Chow emphasizes the urgency of such an approach, as the most vulnerable nuclear command, control, and communications (NC3) satellites will be vulnerable to interference throughout the 2020s.[427] Citing comments from US Air Force Secretary Wilson, Chow noted that the US Space Based Infrared System (SBIRS) is outdated, and vulnerable to attack. While the Air Force is planning a follow-on constellation of new SBIRS satellites in 2029, Chow remains concerned that for the next nine years prior to the update, the existing satellites will remain vulnerable.[428] Even if the US were to race ahead to develop a new, resilient constellation, Chow has expressed skepticism that a truly resilient fleet will be ready in the near or medium-term, and that in the interim, those satellites will need other protections.[429] In addition to being time-intensive, such new, resilient constellations may face other vulnerabilities by nature of being smaller, in addition to having fewer capabilities than their larger counterparts.[430] However, even if such a resilient architecture could

---

425. "Russian Satellite Maneuvers Illustrate Why U.S. Alarm Bells are Ringing," November 6, 2015, https://spacenews.com/editorial-russias-orbital-provocations/
426. Malcolm Davis, "Spy Games in the Grey Zone of Outer Space," ASPI Strategist, February 12, 2020, https://www.aspistrategist.org.au/spy-games-in-the-grey-zone-of-outer-space/.
427. Chow, Brian, "Op-ed: Nuclear vulnerability: In-orbit bodyguards would help protect NC3 satellites from attack," SpaceNews. April 1, 2019. https://spacenews.com/op-ed-nuclear-vulnerability-in-orbit-bodyguards-would-help-protect-nc3-satellites-from-attacks/.
428. Ibid.
429. Chow, Brian, "Stalkers in Space: Defeating the Threat," Strategic Studies Quarterly 11, no. 2 (Summer 2017): pp. 89, https://www.airuniversity.af.edu/Portals/10/SSQ/documents/Volume-11_Issue-2/Chow.pdf.
430. Koplow, "Deterrence as the MacGuffin," 69-71.

come online in time, many worry that a strategy based primarily on protecting arsenals may be inadequate. Such a "deterrence by denial" approach may not stop a determined adversary from continuing to attempt attacks,[431] or from attacking at lower thresholds, such as electronic or cyber interference.[432]

Final agreements about the size and character of warning zones should be discussed and determined by relevant space actors. Chow proposes a 0.2 orbital degree separation for zones, or a zone of approximately 148 km.[433] This standard offers a useful baseline for international negotiations to begin. Within the US, the defense and intelligence communities should consult over establishing a reasonable threshold that would allow the US to effectively protect its satellites without infringing upon the right of states to peacefully move throughout space.

Addressing Simultaneous Tailgating by Multiple Satellites

It should be established that this proposal would apply only to the close, simultaneous approach of satellites towards *multiple* US satellites at any period in time. Individual close approaches to a single US military satellite does not pose as significant a threat, as the US maintains redundant satellite capability to compensate for malfunction, outage, or limited attack.

Simultaneous stalking, however, poses a unique military threat by opening the potential for multiple crucial targets to be destroyed instantaneously. Without classified intelligence, it is not possible for

431. Davis, Paul K., "Toward Theory for Dissuasion (or Deterrence) by Denial: Using Simple Cognitive Models of the Adversary to Inform Strategy," RAND Papers. January 2014. Pp. 3-5 https://www.rand.org/content/dam/rand/pubs/working_papers/WR1000/WR1027/RAND_WR1027.pdf.
432. Lambakis, Steve, "A Guide for Thinking about Space Deterrence and China," National Institute for Public Policy. July 2019. Pp. 4-6 http://www.nipp.org/wp-content/uploads/2019/07/Lambakis-Guide-to-Thinking-About-Space-for-web.pdf.
433. Chow, "Stalkers in Space," 96.

policy advocates to determine how many satellites can be threatened before it becomes a threat to national security, yet international dialogue over this topic can serve as a crucial confidence-building measure between states and demonstrate willingness to admit potential vulnerabilities for the benefit of all actors. As *Bruce MacDonald* describes:

> "While jamming one or two satellites in isolation appears unlikely to quickly escalate into all-out space war (given the longstanding role of electronic warfare in past conflicts), attacking multiple intelligence-gathering satellites would carry a far higher risk of escalation. Somewhere between these two extremes, however, is an uncertain and unknowable boundary that divides offensive space actions that modestly threaten stability from those that are clearly destabilizing and escalatory."[434]

Setting the threshold at one satellite would risk precluding a variety of important space operations. For example, active-debris removal (ADR) operations use advanced RPO capabilities to reliably clean up orbits that have become increasingly congested with debris. Although these operations will prove crucial to prevent space from becoming unsustainably littered with orbital debris, they may also be interpreted as a prelude to an attack given the dual-use nature of space technologies and absence of international dialogue to regulate these technological advancements.[435] Commercial on-orbit servic-

---

434. Bruce MacDonald, "Chapter 2. Space and Escalation" in *Outer Space; Earthly Escalation? Chinese Perspectives on Space Operations and Escalation*, A Strategic Multilayer Assessment (SMA) Periodic Publication, August 2018, https://nsiteam.com/social/wp-content/uploads/2018/08/SMA-White-Paper_Chinese-Persepectives-on-Space_-Aug-2018.pdf.
435. Frank Rose. "The Role of Diplomacy in Keeping Outer Space Safe, Secure, and Sustainable," Remarks at the 32nd Space Symposium, April 14, 2016 https://2009-2017.state.gov/t/avc/rls/255834.htm; Ward Munters and Jan Wouters, "The Road Not Yet Taken for Defusing Conflicts in Active Debris Removal: A Multilateral Organization." Leuven Centre for Global Governance Studies and the Institute for International Law, Working Paper No. 183 – May 2017; Zachary Keck 18, "Space Is Truly the Final Frontier (For the Next Great War)" *The National Interest*, June 17, 2018 https://nationalinterest.org/blog/the-buzz/space-

ing (OOS) spacecraft will also require close approaches to fix and inspect old and malfunctioning satellites, yet these systems can easily be perceived as weapons since they have robotic arms which could be repurposed for offensive operations.[436] Furthermore, space situational awareness operations conducted by the Geosynchronous Space Situational Awareness Program (GSSAP) require close approaches to provide comprehensive knowledge of the space environment and provide states with more reliable pictures of the orbital environment.[437] Considering the inevitable necessity of close-approaches, establishing a threshold of one satellite would reduce the degree of certainty provided by the zone regarding the other party's intent, in turn ratcheting up the risk of misperceptions and inadvertent escalation via incidental close approaches or a failure to enforce the zone due to well-warranted caution.

As a result, the 'warning zone' proposal would apply only to the close, simultaneous approach of satellites towards *multiple* US satellites at any period in time. In part, warning zones are functionally and legally distinguished from KOZs due to this caveat. A state does not claim sovereignty over, nor will it preclude others from entering, an area; free passage will be permitted, yet states will have a clearer understanding when passage through the warning zone is conducted for peaceful purposes (such as ADR and OOS missions), or whether passage should be interpreted as hostile intent.

---

truly-the-final-frontier-the-next-great-war-26284.
436. Joshua P. Davis, John P. Mayberry, and Jay P. Penn, "ON-ORBIT SERVICING: INSPECTION, REPAIR, REFUEL, UPGRADE, AND ASSEMBLY OF SATELLITES IN SPACE," CENTER FOR SPACE POLICY AND STRATEGY, April 2019, https://aerospace.org/sites/default/files/2019-05/Davis-Mayberry-Penn_OOS_04242019.pdf.
437. Steven Lambakis, Foreign space capabilities: Implications for U.S. national security, *Comparative Strategy* 37(2): 115, 2018, https://www.tandfonline.com/doi/full/10.1080/01495933.2018.1459144?scroll=to; Michael Nayak, "CubeSat proximity operations: The natural evolution of defensive space control into a deterrence initiative," *The Space Review*, 2016, http://www.thespacereview.com/article/2902/1.

The Proposal Only Applies to Geosynchronous Orbit

International discussions should initially only focus on establishing warning zones in geosynchronous orbit (GEO). First, orbital physics makes this proposal easier to implement in GEO. Satellites in GEO, which rotate at the same velocity as the planet, are essentially parked above the earth (whereas satellites in LEO and MEO have far more complicated and occasionally overlapping orbital trajectories).[438] Due to complex satellite trajectories in the LEO, which will be magnified by increased commercialization, the LEO is "highly congested and only small zones would be feasible," meaning that warning zones would offer little protection because it would be far more difficult to ascertain another state's intent when positioning satellites near other satellites.[439] Transversal of a warning zone in GEO could be clearly construed as conveying hostile intent, as no state would have a legitimate peaceful motivation to position satellites in close proximity to another state's military and intelligence satellites in GEO.

Second, GEO is home to many vital and vulnerable national security satellites. The majority of satellites in LEO are commercial and are less likely to direct American military and intelligence operations. It would be disproportionate to construe a close approach to a commercial or weather satellite as constituting aggression or a prelude to an armed attack. Furthermore, many national security satellites in GEO, such as NC3 satellites, are old, fragile, and incapable of being maneuvered out of the way of close approaching objects. Adversaries would likely target these NC3 satellites at the outbreak of a conflict as a prelude to a first strike, making it critical that they remain defended and protected. Furthermore, a state fearing an adversary's disabling attack on NC3 satellites may strike first rather than lose their nuclear deterrent capability. Warning zones around these satellites would allow states to ascertain whether adversaries are about to

438. David Wright, Laura Grego, and Lisbeth Gronlund, "The Physics of Space Security: A Reference Manual," (Cambridge, MA: *American Academy of Arts & Science*, 2005), https://aerospace.csis.org/wp-content/uploads/2019/06/physics-space-security.pdf.
439. Newsome, "The Legality of Safety and Security Zones in Outer Space," 17.

execute a first strike, reducing the risks of miscalculation.[440]

Third, an international space traffic management system for assigning orbital slots in GEO already exists. The ITU assigns orbital slots of an average of 2 degree orbital separation, far larger than the warning zones suggested by this proposal.[441] More specifically, to comply with these warning zones, states "need not reposition any of their operational satellites to observe the above suggested rule of 0.2-degree minimum satellite separation between any pair of... GEO satellites."[442] Applying and enforcing zones in LEO represents far more of a challenge because of commercial sector involvement. Chow acknowledges potential difficulties in asking commercial providers to monitor and change orbital paths to avoid traversing zones.[443] Furthermore, commercial space firms and consultancies have expressed concern about the changes in trajectories required to avoid zones. Scott Kordella, the director of outer space systems at Mitre, fears that extensive new US licensing regimes in space could sacrifice American commercial leadership.[444] In the status quo, most spacecraft in LEO lack transponders and specific orbital slots; the creation and enforcement of specific orbital slots (and warning zones surrounding those slots) would require the creation of a licensing regime.[445] Others have argued that maneuvering in LEO

440. Brian Chow. "Op-Ed | Nuclear Vulnerability: In-Orbit Bodyguards Would Help Protect NC3 Satellites from Attacks," SpaceNews.com, April 1, 2019, https://spacenews.com/op-ed-nuclear-vulnerability-in-orbit-bodyguards-would-help-protect-nc3-satellites-from-attacks/.

441. "Sharing the Sky - ITU's Role in Managing Satellite and Orbit Spectrum Resources" ITU Background, Plenipotentiary 2014, https://www.itu.int/en/plenipotentiary/2014/newsroom/Documents/backgrounders/pp14-backgrounder-sharing-the-sky.pdf.

442. Chow, "Stalkers in Space," 96.

443. Ibid., 94-97.

444. Kordella, Scott, "Op-ed: Protecting low Earth orbit from becoming the new Wild West," *Space News*. March 16, 2019. https://spacenews.com/op-ed-protecting-low-earth-orbit-from-becoming-the-new-wild-west/.

445. LaunchSpace, an educational and consulting organization dedicated to training and continuing education for space professionals and to supporting the aerospace community, "Not A Job, But An Adventure - Space Traffic Controller," 11 September 2018, Space Daily, http://www.spacedaily.com/reports/

might be frequent enough to require burdensome "active guidance measures" for each commercial operator.[446] The risk is particularly salient as onerous launch and licensing restrictions might hamper further commercial space development.[447]

However, international deliberation could eventually expand to include the development of warning zones in LEO, and proposals could involve negotiations with the private sector. Zones could also be integrated into a broader space traffic management regime, yet such a system will require extensive cooperation with the private sector and all spacefaring states. Similarly, warning zones in GEO would be ideally established by consensus in a multilateral forum, but they can also be agreed to on bilateral bases that would still provide mutual understanding and reduce the risk of misperception.[448]

Not_A_Job_But_An_Adventure___Space_Traffic_Controller_999.html.

446. Collins Patrick, "Legal Considerations for Traffic Systems in Near-Earth Space." Proceedings 31st Colloquium on the Law of Outer Space, IISL, pp. 296-303. http://www.spacefuture.com/archive/legal_considerations_for_traffic_systems_in_near_earth_space.shtml.

447. Joshua Hampson, "The Future of Space Commercialization," Niskanen Center, January 25, 2017. Pp. 14-19, 31-34 https://niskanencenter.org/wp-content/uploads/2017/01/TheFutureofSpaceCommercializationFinal.pdf.

448. For more comprehensive versions of this proposal that would still apply in LEO see Brian G Chow, "Space Arms Control: A Hybrid Approach," Strategic Studies Quarterly 12, no. 2 (Summer 2018): 107-132, https://www.airuniversity.af.edu/Portals/10/SSQ/documents/Volume-12_Issue-2/Chow.pdf; Geoffrey Forden "After China's Test: Time for a Limited Ban on Anti- Satellite Weapons," Arms Control Today , vol. 37 (April), https://www.armscontrol.org/act/2007-04/features/after-chinas-test-time-limited-ban-anti-satellite-weapons; James Alver, Andrew Garza, Christopher May, "An Analysis of the Potential Misuse of Active Debris Removal, On-Orbit Servicing, and Rendezvous & Proximity Operations Technologies," 5-6-19, Secure World Foundation, https://swfound.org/media/206800/misuse_commercial_adr_oos_jul2019.pdf, Mark A. Gubrud, "Chinese and US Kinetic Energy Space Weapons and Arms Control," *Asian Perspective* 35, no. 4 (2011): 628-631.

Cooperative Approaches

To establish warning zones, the US should adopt a cooperative posture. The current absence of international negotiations over space stability have created a culture of distrust and hostility that risks increasing tensions which could spiral into conflict.[449] In particular, the US has damaged its international credibility over space arms control by consistently vetoing treaties such as the Treaty on the Prevention of the Placement of Weapons in Outer Space, the Threat or Use of Force Against Outer Space Objects (PPWT) and the Prevention of an Arms Race in Space Treaty (PAROS). Proposing negotiations on the basis of peaceful intent will help reverse the perception that the US aims to aggressively dominate the space domain.[450]

The US must be wary of the prospect that other states, particularly Russia and China, would view any arms control proposal as a veil for US militarization of outer space. Even Chow admits that "[r]ealistically, potential adversaries are far more likely to want the United Nations to adopt alternative space proposals, which ignore the space-stalking threat."[451] However, the UN thus far has failed to generate proposals that overcome definitional hurdles and enforcement-related barriers. The PAROS and the PPWT proposed and supported by Russia and China both neglected to define "space weapons," and would likely not include any restrictions on co-orbital ASAT capabilities as they are not designed to conduct space-to-ground strikes. With regards to the PPWT, concerns have also been raised that China may be negotiating in bad faith to insulate their space arsenal from regulation: the PPWT would allow ground-to-space (direct-ascent)

---

449. Brian Weeden and Xiao He, "U.S.-China Strategic Relations in Space" *The National Bureau of Asian Research*, NBR Special Report #57, April 2016, https://www.nbr.org/wp-content/uploads/pdfs/publications/special_report_57_us-china_april2016.pdf; "Attacking Satellites Is Increasingly Attractive—and Dangerous," *The Economist*, July 18, 2019, https://www.economist.com/briefing/2019/07/18/attacking-satellites-is-increasingly-attractive-and-dangerous.
450. Michael Listner, "The art of lawfare and the real war in outer space," *The Space Review*, September 17, 2018, http://www.thespacereview.com/article/3571/1.
451. Chow, "Stalkers in Space," 109.

ASAT capabilities, of which China possesses a veritable suite. Russian and Chinese opposition to the International Code of Conduct for Outer Space Activities proposed by the European Union, which would have included a right to preemptive self-defense, underscores the likelihood that Russia and China would likely refuse to participate in accords to cooperatively establish warning zones.[452]

The US should begin international negotiations over warning zones before declaring them unilaterally. Declaring zones without prior deliberation might leader other nations to perceive these zones as a guise for militarizing space. Other states may also probe the credibility of these declared zones, particularly to exploit a lack of international consensus on their legality. Although Russia or China might violate unilaterally declared zones, the US should be careful to respond absent certainty that preemptive self-defense is legal. Opening negotiations to all relevant parties, adversaries and allies included, will give states the opportunity to air grievances regarding warning zones before they are established. Even if Russia or China ultimately reject the proposal, the US will be able to point to its good faith effort of initiating dialogue and respecting Chinese and Russian concerns. As Chow describes:

> "In the worst-case scenario where no space arms control comes out of this hybrid approach, the United States would still have acted in good faith for pursuing an international space arms control treaty. Consequently, the world would have more understanding and support toward the United States as it had no choice but to switch to unilateral space arms control measures to ensure space security and stability."[453]

---

452. *The Economist*, "Attacking satellites is increasingly attractive—and dangerous," July 18, 2019. https://www.economist.com/briefing/2019/07/18/attacking-satellites-is-increasingly-attractive-and-dangerous; James Alver, Andrew Garza, Christopher May, "An Analysis of the Potential Misuse of Active Debris Removal, On-Orbit Servicing, and Rendezvous & Proximity Operations Technologies," Secure World Foundation, May 6, 2019. https://swfound.org/media/206800/misuse_commercial_adr_oos_jul2019.pdf, 15-16.
453. Brian G Chow, "Space Arms Control: A Hybrid Approach," Strategic Stud-

While there will inevitably be detractors, zones will be most stable and effective when they receive international buy-in.

There is some evidence to suggest that Russia and China would also accept the warning zone proposal. First, Russian and China might recognize that warning zones are mutually beneficial, giving them a legal right to defend their satellites while maintaining stability as a variety of RPO capable spacecraft are rapidly deployed.[454] Even if the two parties do not accede to the agreement at first, the failure of existing proposals to address dual-use threats from RPO capable spacecraft might encourage Russia and China towards acceptance of warning zones.

Second, the United States can gain political leverage and engage in "full-contact lawfare" by highlighting the contradictory positions of Russia and China regarding space arms-control diplomacy.[455] As the two nations continue to conduct potentially destabilizing RPO tests in orbit, the international community will eventually recognize their dual-use nature as ASATs. This is especially salient considering the recent Russian testing of the Nudol, as well as the much-pilloried Chinese direct-ascent and debris-producing ASAT test of 2010.

With regards to China, this reality is particularly relevant when one considers China's desire to leave a mark on international institutions in the space domain. For example, Brian Weeden and Xiao He, an assistant research fellow at the Institute of World Economics and Politics in the Chinese Academy of Social Sciences, argue that China feels boxed out of most international arenas where the US is already well-established. Space offers an area where China can proactively participate and lead in international governance, shaping rules to align with their own interests.[456] Furthermore, He writes that

---

ies Quarterly 12, no. 2 (Summer 2018): https://www.airuniversity.af.edu/Portals/10/SSQ/documents/Volume-12_Issue-2/Chow.pdf. 129.

454. Chow, "Space Arms Control: A Hybrid Approach," 115-119.

455. Listner, "The art of lawfare and the real war in outer space."

456. Xiao He and Brian Weeden, "U.S.-China Strategic Relations in Space," The National Bureau of Asian Research. April 2016. Pp. 73-74. https://www.nbr.org/wp-content/uploads/pdfs/publications/special_report_57_us-china_april2016.

China has recognized international resistance to the PPWT proposal and considers the likelihood of success as remote. Furthermore, Chinese President Xi Jingping has emphasized cooperative security a central aspect of China's foreign political approach, hoping to produce win-win outcomes in contrast to the zero-sum thinking that dominated during the Cold War.[457] Additionally, China has an economic incentive to pursue stability in orbit. As China continues to use satellites to expand its Belt-and-Road Initiative (BRI), they have a strong incentive to protect their own assets.[458]

Such international and economic incentives similarly exist for Russia, a declining power with a desire to increase international engagement in orbit. For Russia, space has long been an area of great national pride. In recent years, however, scientific projects and joint ventures in space have declined alongside a steadily weakened economy.[459] Hoping to maintain its status as a great power in space, Russia would likely be interested in the prestige associated with a major cooperative venture.[460] While warning zones are not a form of scientific cooperation, the process of negotiating new, groundbreaking norms for outer space gives Russia a much-desired seat at the table in international negotiations.[461] While China and Russia may prefer to wait for their preferred approach to arms control, if the US holds firm and makes clear that it will not accede to the PPWT, there may be increasing willingness on the part of China to return to the negotiating table on American terms in the face of pressure.

---

pdf.

457. Sverre Lodgaard, "Arms Control and World Order," Journal for Peace and Nuclear Disarmament, Vol. 2. Issue 1. Pp. 17-19. https://www.tandfonline.com/doi/full/10.1080/25751654.2019.1631243.

458. Brian Chow and Henry Sokolski, "Priority-One for Space Policy Should Be to Protect U.S. Satellites," SpaceNews.com, October 12, 2019, https://spacenews.com/op-ed-priority-one-for-space-policy-should-be-to-protect-u-s-satellites/.

459. Pavel Luzin, "Russia's Position in Space," Foreign Affairs, September 21, 2016. https://www.foreignaffairs.com/articles/2016-09-21/russias-position-space.

460. Ibid.

461. Nicole Jackson, "Outer Space in Russia's Security Strategy," Simons Paper in Security and Development. No. 64, August 2018. 5.

Potential Objections

There exist a number of potential objections to the warning zones proposal. This section aims to distill the most relevant objections and to demonstrate why many of these objections reinforce the necessity of establishing warning zones in outer space.

Space Situational Awareness (SSA)

Military and intelligence officers may worry that restrictions on RPO behavior may prevent the US from conducting close approaches with surveillance satellites, such as through the Geosynchronous Space Situational Awareness Program (GSSAP). Described as the "neighborhood watch," GSSAP satellites are considered crucial for deterring adversary aggression by signaling the United States is continually monitoring activities on orbit.[462] GSSAP satellites are capable of performing RPO operations and occasionally make close approaches to adversary satellites on surveillance missions or to signal that the United States is constantly watching. The proposed warning zones in this approach would not impede operations conducted by GSSAP satellites. Individual close approaches for surveillance purposes would not pose a military threat and would not rise to the level of demonstrating hostile intent. As of now, the United States does not field capabilities to allow multiple simultaneous close approaches and would sacrifice little by agreeing to a mutual restriction on behavior in space. Furthermore, multiple close approaches could be conducted over multiple days instead of simultaneously.[463]

Furthermore, some have argued that monitoring and enforcing

---

462. Steven Lambakis, Foreign space capabilities: Implications for U.S. national security, *Comparative Strategy* 37, no. 2 (2018): 115, https://www.tandfonline.com/doi/full/10.1080/01495933.2018.1459144?scroll=to; Michael Nayak, "CubeSat proximity operations: The natural evolution of defensive space control into a deterrence initiative," *The Space Review*, 2016, http://www.thespacereview.com/article/2902/1.
463. Chow and Sokolski, "Op-Ed | Priority-One for Space Policy Should Be to Protect U.S. Satellites."

zones would be too burdensome; keeping tabs on every satellite in orbit would be a challenge that is, in the words of Steven Lambakis, "monumental."[464] However, the warning zones proposal is expressly limited in scope to activities in GEO, as opposed to the substantially more difficult task of monitoring in LEO.[465] In GEO, existing US space situational awareness (SSA) capabilities may fill the gap. Chow notes that the US Geosynchronous Space Situational Awareness Program (GSSAP) already performs the role of monitoring activities in GEO, with sufficient precision and timeliness to alert the DOD of potential zone breaches.[466]

Loitering

Zone opponents frequently argue that they provide limited safety, as adversaries could easily park satellites outside of the zones and still be prepared to execute a devastating attack.[467] Concerns about loitering merit attention, yet they do not obviate the importance of the zone. If a determined adversary wants to attack a US satellite during a crisis, they will inevitably do so. Warning zones are not meant to provide absolute protection to satellites, but instead will help clearly define violations as constituting hostile intent. In the current environment, adversaries can loiter as close to a satellite as they wish. Warning zones will enable states to easily ascertain intent from adversaries that loiter outside of, or enter, a zone. SDZs and KOZs offer the illusion of protection for satellites, yet they cannot reliably defend against all incoming attackers.

---

464. Lambakis, Steven, "On the Edge of Earth: The Future of American Space Power," The University Press of Kentucky. Pp. 120-130 2001.

465. Chow "Stalkers in Space," 98.

466. Ibid.

467. Newsome, "The Legality of Safety and Security Zones in Outer Space,"; Alex Gliksman, "Options for Space Arms Control," *America Plans for Space,* A Reader Based on the National Defense University Press, Dec 15, 1987, https://apps.dtic.mil/dtic/tr/fulltext/u2/a187465.pdf; James A. Vedda and Peter L. Hays, "Major Policy Issues in Evolving Global Space Operations." The Mitchell Institute for Aerospace Studies, Air Force Association. https://tinyurl.com/y4cj6ran.

Deterrence

Some believe the United States should reject limitations on behavior in space and strive for complete freedom of action in orbit.[468] Advocates of offensive space control argue that the United States can only achieve peace through strength and should reject any constraints on military capabilities in the space domain. Many of these detractors fear that China or Russia will inevitably cheat on any mutual restrictions to leapfrog US capabilities and erode US military advantage. Many of these ideas rest on faulty assumptions, yet even proponents of offensive dominance will be in favor of establishing warning zones around vulnerable satellites.

First, establishing warning zones would not prevent the United States from using RPO operations to hold adversary satellites at risk. The United States could still conduct counter space missions; adversaries would gain clarity that these close approaches are undertaken to convey hostile intent. Crisis bargaining would become far more stable and predictable if the international community were to agree upon warning zones (and how they can be peacefully traversed), otherwise close approaches could always be perceived as a likely prelude to war.

Second, few have yet to articulate a reason why the United States needs to conduct multiple close approaches simultaneously to execute a counter space mission. The United States can develop a slate of counter space capabilities including direct-ascent ASATs,

468. US National Space Policy, 2006, https://aerospace.org/sites/default/files/policy_archives/Natl%20Space%20Policy%20fact%20sheet%2031Aug06.pdf; Christopher Stone, "Re-thinking the National Security Space Strategy: Chinese vs. American perceptions of space deterrence". The Space Review. 2013. http://www.thespacereview.com/article/2395/1; Christopher Michael Stone, "Defending critical space infrastructure," Space News, December 5th, 2016 http://www.spacenewsmag.com/commentary/defending-critical-space-infrastructure/; Baker Spring and Michaela Bendikova, "More Limits on U.S. Space Systems Unacceptable," http://thf_media.s3.amazonaws.com/2012/pdf/wm3470.pdf; John Yoo and Jeremy Rabkin, "Chapter 7: Coercion in Space," in "Striking Power: How Cyber, Robots, and Space Weapons Change the Rules for War,". Encounter Books. 2017.

cyber weapons, and jammers that can hold adversary satellites at risk without violating warning zones. The United States can also continue to test RPO capabilities through limited close approaches, so the United States can still use these systems for offensive missions if needed.

Third, most RPO capable satellites are better suited for defensive space operations than for asserting offensive control. Some of the most likely systems, small satellites known as Cubesats, are ideal defensive systems because they can act as bodyguards to protect otherwise vulnerable satellites.[469] As space is an offense-dominant domain and the United States possesses an asymmetric dependence on space assets for military operations, it would be preferable for the United States to primarily concentrate on defending satellites before attempting to dominate the space domain.

Boundary Testing

Critics of the proposed warning zones may fear that establishing these zones would encourage further boundary testing by adversaries. Publicly declaring a zone may encourage adversaries to probe American resolve to see if the United States will actually punish violations or prefer to avoid confrontation in space. However, maintaining strategic ambiguity over how the United States may punish violations could create a superior deterrent effect for adversaries may fear the consequences of American retaliation. A public declaration that the United States will not tolerate close approaches without a specified response option may leave the United States with

469. Michael Nayak, "CubeSat proximity operations: The natural evolution of defensive space control into a deterrence initiative," 2016, http://www.thespacereview.com/article/2902/1; Edward Ferguson and John Klein, "The future of war in space is defensive," The Space Review, December 19, 2016, https://www.thespacereview.com/article/3131/1; Michael Nayak, "Deterring Aggressive Space Actions with Cube Satellite Proximity Operations: A New Frontier in Defensive Space Control," *Air and Space Power Journal* 31, Vol. 4, pp. 92-103, https://www.airuniversity.af.edu/Portals/10/ASPJ/journals/Volume-31_Issue-4/SEW-Nayak.pdf.

extra flexibility to punish violations as seen fit.

These detractors overlook some central concerns. First, adversaries are already exploiting the lack of international agreements to continue grey zone provocations.[470] Adversaries are far more likely to probe US boundaries in the absence of international consensus that stalking should be considered hostile. Russia and China, in particular, will no longer be able to exploit the lack of international consensus against stalking if they are invited to international negotiations. Even if they reject the proposed deal, the rest of the international community might condemn their boundary testing and recognize an American right to preemptive self-defense if necessary.

Second, threats of asymmetric retaliation to violations of warning zones will ring hollow and not effectively deter adversary aggression. Adversaries will recognize American threats as illegal, disproportionate escalation.[471] In particular, threats to retaliate terrestrially will be seen as unnecessarily expanding the domain of conflict, ensuring that the international community does not support the American response. American dependence on space also makes these disproportionate threats ineffective, as adversaries will always have more to gain by destroying American space capabilities than they have to lose.

Conclusion

Governments and militaries have little experience discerning intent in space operations given a lack of crisis management experience in orbit. Warning zones have the potential to enhance mutual understanding and allow states a means to peacefully signal intent inside

---

470. Malcolm Davis, "Spy Games in the Grey Zone of Outer Space," ASPI Strategist, February 12, 2020, https://www.aspistrategist.org.au/spy-games-in-the-grey-zone-of-outer-space/.
471. See Koplow, David A. 2019. Deterrence as the MacGuffin: The Case for Arms Control in Outer Space. Rochester, NY: Social Science Research Network. SSRN Scholarly Paper. http://papers.ssrn.com/abstract=3436311 (May 25, 2020).

and outside of crises. Agreed upon in an amicable and cooperative setting, norms of behavior in space offer a crucial mechanism for states to credibly signal their intent during crises. With the contemporary lack of international agreement regarding co-orbital activities, a state may preposition a satellite during a crisis, yet it is unclear that their adversary will receive the intended signal. By agreeing to this warning zone proposal, states can reliably interpret when violations of zones are peaceful or hostile. This ensures security for defending states by giving them a signal of impending adversary aggression and allows aggressing states to ensure their signals of hostile intent are properly interpreted during crises.[472]

472. Audrey Schaffer, "The Role of Space Norms in Protection and Defense," JFQ 87, 4th Quarter 2017, 88-92, https://ndupress.ndu.edu/Portals/68/Documents/jfq/jfq-87/jfq-87.pdf.

Chapter 6:

Missile Wars: What Awaits

Henry D. Sokolski

Now that the United States and Russia have agreed to extend New START for five more years, the question arises, what comes next. The Russians have evinced a desire to revive the Open Skies agreement and to agree to a moratorium on the deployment of intermediate range nuclear weapons. They also wish to revive the 2015 Iran nuclear deal and have long been pushing for an international agreement on certain norms for space-related activities.[473]

Realizing these Russian diplomatic possibilities, however, are unlikely to be as important as what, if anything, Washington and Moscow might do either to counter or to limit China's growing strategic military capabilities. State Department officials are hopeful that agreement with Russia on New START will ease the way to getting Beijing to agree to limit its strategic weaponry. But assuming this is possible, what would any arms limitation deal look like and when might it be reached?

The specifics are unclear. What would drive such talks are not. They will reflect a long-term missile-driven military competition that has already begun. At the high end, this long-term military rivalry involves not just the United States, but America's East Asian and European allies, and, to a lesser extent, Russia. It will create ample

473. See, Ben Aris in Berlin, "The Kremlin launches drive to revive Cold War missile treaties nixed by US," *Intellinews*, January 28,2021, available at http://www.intellinews.com/the-kremlin-launches-drive-to-revive-cold-war-missile-treaties-nixed-by-us-201519/ .

opportunities for diplomacy as well as for increased military competitions and possible miscalculations. At the mid and low-end, there also will be military and diplomatic contests driven by the spread of accurate missiles, drones, and related technologies to smaller states and non-state actors. The top missile rivalry, however, will be with the Chinese. This missile-driven competition between the United States and its allies with China will take decades to unfold, and will come in stages, most likely along the following lines:

2021-2026

During the early years of the missile-driven rivalry with China (approximately 2021 through 2025), the United States will be able to strike China's homeland with many more long-range missiles than China will be able to fire against the continental United States. Initially, all of these intercontinental-range missiles, on both sides, will be nuclear-armed. From now through the mid-2020s, though, China will have far more nonnuclear short, medium, and intermediate range missiles and drones in East Asia than the United States or its allies will. American and Chinese active and passive missile defenses during this phase will continue to be less than perfect but effective enough to force both sides to increase the size of their offensive missile forces to be able to destroy any given target. China, as a result, will oppose all forms of US and allied active missile defense.

Meanwhile, the United States and its allies will continue to rely heavily on space-based systems to meet their military command, communication, control, intelligence, and surveillance requirements. America's space-based military systems will be essential to its conduct of defensive and offensive missile operations. China and Russia will rely far less on their space-based systems to achieve their military goals. Instead, they will focus their military space efforts on neutralizing US and allied military space assets.

For most of this period, US and allied space-based systems will be vulnerable to Russian and Chinese ground-based lasers. These systems can dazzle and damage allied satellite optical sensors and by the mid to late 2020s will be able to disable the satellites themselves, destroying their metal skins. During this same period, rendezvous satellites, designed to refuel, repair, and reposition spacecraft into their desired orbits, could also be used as anti-satellite weapons to de-fuel, damage, and disable them by pulling them out of position. Neither of these anti-satellite systems nor high-power electronic anti-satellite warfare systems would necessarily produce any space debris.

2026-2031

By the end of the decade, China will begin to develop and deploy long-range, accurate, nonnuclear missiles that could target the continental United States with maneuverable re-entry vehicles and hypersonic technologies. China would continue to use intermediate, medium, and shorter-range versions of such missiles to threaten America's regional friends and allies. By 2026, the United States and its East Asian allies are likely to begin closing this missile gap with some number of short, medium, and intermediate range nonnuclear missile systems based in South Korea, Japan, US territories, and perhaps Taiwan, Vietnam, Australia, and the Philippines. Also, by the mid to late -2020s, both China and the United States will begin to deploy advanced bombers capable of delivering large numbers ("swarms") of nonnuclear missiles and drones.

All of these offensive developments, in turn, will increase US and allied interest in developing enhanced passive and active missile defenses, including ultra-performance concrete protected structures, high mobility offensive and defensive missile launchers, and advanced decoys and camouflage. In general, a premium would be placed on developing a variety of high-fire rate, low-cost-per-shot active defenses, and using passive defenses to increase the number of missiles that the Chinese would have to fire to destroy any given US or allied target.[474] Missile defense cooperation among

474. See Bryan Clark and Henry Sokolski, "Thinking Hard about Missile Defense," *Real Clear Defense*, August 5, 2020, available at https://www.realcleardefense.com/articles/2020/08/05/thinking_hard_about_missile_

states friendly with the United States and NATO Nations is likely to increase. Missile defense coalitions might also see advantage in coordinating their missile defense and missile offensive technology exports and exchanging intelligence on the missile programs of common adversaries.

The United States and its allies could further seek to put China on the defensive by targeting assets essential to the Chinese Communist Party's (CCP's) rule over provinces that might otherwise go their own way. The aim would be to deter Chinese adventurism and to encourage the CCP to invest much more in active and passive defenses on the mainland. These Chinese defenses would protect domestic targets that might otherwise go undefended and come at the cost of China investing more in offensive strike systems.

Against such US and allied efforts, Beijing might hedge its strategic bets. It might develop a large "peaceful" nuclear fissile material production base, one Beijing could ramp-up relatively quickly to produce large numbers of nuclear weapons, if it chose, by 2030, by exploiting its civilian nuclear reprocessing and fast reactor plants as well as its massive uranium enrichment capacity.[475]

China might also redouble its anti-satellite programs as well as its efforts to explore and command deep space regions near and beyond the moon. Beijing might also increase strategic uncertainties by continuing to collaborate with (a declining) Russia on a variety of strategic weapons programs and joint training operations. These Chinese hedges would likely increase American interest in developing low-vulnerability space-based command, control, communication, and surveillance systems as well as missile defense, anti-satellite lasers, bodyguard satellites, and other active and passive satellite defense systems.

---

defense_115519-full.html.

475. See Henry Sokolski, editor,China's civil nuclear sector: swords from Plowshares?Arlington, VA: The Nonproliferation Policy Education Center, Occasional Paper 2021 03, March 2021).

2031-2040

Once both the United States and China have larger numbers of long and shorter range precise strike systems and much more robust active and passive missile and air defenses, the possibility of major missile exchanges would be far less attractive to both sides than today. China, during this period, would likely have achieved nuclear parity with the United States and Russia. This would likely make the early use of nuclear weapons more remote.[476]

With considerable effort, the United States could have much more resilient and survivable space-based systems to support its military. The precise targeting of stationary military targets, and mobile ones, to some extent, would be likely, as would the targeting of political locations and facilities that China views as necessary to maintain political control. This may also be true of Chinese targeting of Taiwan and Japan, and to a lesser extent, of the United States.

In this regard, America might enjoy some advantage: Unlike China, the United States government should enjoy more popular domestic support than the Chinese Communist Party. America should, therefore, be politically more resilient against attacks against its political nodes of power. Also, China has some of the world's largest concentrations of population in some of the world's largest cities. This could make it more difficult for China to target a "significant" number of commercial-political nodes in the United States without risking more costly counter strikes from the United States.

Plus Other Not-So-Great-Power Missile Competitions

In addition to contemplating future missile competitions with Moscow or Beijing, the United States and its friends will also have to contend with the spread of advanced, accurate missile systems and

476. See Henry Sokolski, "Hiroshima at 75: A Peek into Our Nuclear Future," *The American Interest,* August 6, 2020, available at https://www.the-american-interest.com/2020/08/05/a-peek-into-our-nuclear-future/ .

drones to medium powers, weak states, to sub-nationals. These actors' acquisition and possible use of these weapons could be used to catalyze wars.

Hezbollah's, Houthi, Iranian, Saudi, Turkish, UAE Israeli, Pakistani, Indian, South and North Korean and Taiwanese military forces all are developing and deploying such precision missiles. Any one of them may try to deter or to escalate conflicts by threatening value targets (not just population centers, but sympathetic systems including reactors, petrochemical plants, natural gas depots, dams, desalination units, electrical, water distribution systems, and communication nodes) with accurate missiles or drones. Such strikes, in turn, could escalate into much larger conflicts, which could conceivably engage nuclear-armed states, including China, Russia, Pakistan, India, Israel, the United States or NATO.

Three recent developments are suggestive:

September 14, 2019 a missile attack against the oil facilities at Abqaiq in Saudi Arabia[477] penetrated three missile defense systems. The attacking missiles hit their targets with pinpoint accuracies. This was arguably the first time highly accurate missiles were aimed at point targets protected by several layers of air and missile defenses (none of which were engaged).

August 4, 2020, an ammonium nitrate explosion 0.5 to 1.1 kiloton in yield was set off in Beirut.[478] Unlike other previous, massive, chemical explosions, this one was televised internationally. As such, it established a new normal for high-end urban eruptive disruptions.

July 15, 2020, an Azerbaijani defense ministry spokesman threatened Armenia with accurate missiles strike against its Metsamor nuclear power plant.[479] This was the first time a government publicly

---

477. *Wikipedia*, "2019 Abqaiq–Khurais Attack," November 30, 2020, Available at https://en.m.wikipedia.org/wiki/2019_Abqaiq%E2%80%93Khurais_attack.
478. *Wikipedia,* " 2020 Beirut Explosion, " December 4, 2020, available at https://en.m.wikipedia.org/wiki/2020_Beirut_explosion.
479. Vladimir Vardanyan, "Azerbaijan Threats to attack the Armenian Nuclear Power Plant in Violation of the International Law," *Armenpress*, July 15, 2020,

threatened to attack a civilian reactor with accurate missiles. Shortly thereafter, Russian President Putin called Turkish President Erdogan to join him in "stabilizing" the crisis.[480]

Previously, Armenian and Azeri officials threatened to attack other value targets with missiles including petrochemical plants, dams, and cities.[481] In 2020, both Armenia and Azerbaijan used drones and missiles against civilians. They used these same systems for pin-point strikes against military assets.

These events should be viewed in the context of the dramatic spread of increasingly accurate, ever longer range cruise and ballistic missiles and drones to weak states and state-sponsored proxies, such as the Houthis[482] and Hezbollah.[483]

Meanwhile, the commercial use and availability of high performance and heavy lift drones is increasing. Drones already are being tested using artificial intelligence to help guide them without GPS signals.[484] Drones also have been mated with facial recognition,

---

available at https://armenpress.am/eng/news/1022038.html.

480. Reuters, "Russia's Putin, Turkey's Erdogan Discuss Conflict Between Armenia and Azerbaijan," *US News and World Report,* July, 27, 2020, available at https://www.usnews.com/news/world/articles/2020-07-27/russias-putin-turkeys-erdogan-discuss-conflict-between-armenia-and-azerbaijan.

481. Paul Goble, "Armenian Nuclear Power Plant Able to Withstand Attack, Russian Security Expert Claims," *Jamestown Monitor*, July 21, 2020, available at https://jamestown.org/program/armenian-nuclear-power-plant-able-to-withstand-attack-russian-security-expert-claims/ and Trend, "Azerbaijan sends letter to UN Secretary General regarding Armenia's ongoing aggression," *AZERNEWS,* August 7, 2020, available at https://www.azernews.az/karabakh/167792.html.

482. Ian Williams and Shaan Shaikh, "Report: The Missile War in Yemen," Missile Threat, Center for Strategic and International Studies, June 9, 2020, last modified July 8, 2020, https://missilethreat.csis.org/report-the-missile-war-in-yemen/.

483. Shaan Shaikh, "Missiles and Rockets of Hezbollah," *Missile Threat,* Center for Strategic and International Studies, June 26, 2018, last modified September 27, 2019, https://missilethreat.csis.org/country/hezbollahs-rocket-arsenal/ .

484. "Israel carries out first-ever worldwide drone test without GPS, succeeds," *The Jerusalem Post,* February 14, 2020, available at https://www.jpost.com/

presaging nightmare "slaughterbot" scenarios of massive, selective assassinations.[485] On the flip side, microwave defense systems are proving effective against drones and may well afford some protection against faster flying precision missiles.[486]

Diplomatic Implications and Opportunities

Regarding the spread of accurate missiles and drones to relatively weak states and non-state actors, the major missile-supplying states will first try to develop their own active and passive defenses. States friendly with one another bilaterally and multilaterally (e.g., NATO, ANZUS, ASEAN, the Quad, Abraham Accord states, etc.), however, will also likely cooperate on such defenses, sharing threat assessments, defense systems, and training.

This cooperation, in turn, could be exploited to encourage the adoption of shared diplomatic missile control narratives. This could include agreed positions on amending the Missile Technology Control Regime to cover new, critical missile technologies that apply to systems beneath the regime's current nuclear-capable range-payload criteria. It also could include coordination of missile technology export controls and intelligence sharing among friendly states and collections of states (e.g., again, NATO, ANZUS, ASEAN, the Quad, Abraham Accord states, etc.).

These shared diplomatic missile control narratives could also factor in missile diplomacy with the world's major missile rivals — Russia, China, and the United States. Initially, diplomatic missile overtures might be modest. Certainly, with China's regional missile

---

jpost-tech/israel-carries-out-first-ever-worldwide-drone-test-without-gps-succeeds-658886.

485. See StratoEnergics, "Slaughterbots," video 2017, available at https://www.youtube.com/watch?v=9CO6M2HsoIA&feature=youtu.be .

486. Joseph Trevithick, "Air Force Set To Deploy Its Counter-Drone "Phaser" Microwave Weapon Overseas," *The Drive*, September 24, 2019, available at https://www.thedrive.com/the-war-zone/29992/air-force-set-to-field-test-its-counter-drone-phaser-microwave-weapon-overseas-in-2020.

advantage and America's intercontinental nuclear superiority for the next five to 10 years, Washington and its East Asian allies would have difficulty persuading China to make arms control deals of the counting-numbers sort previously reached with the Russians. With this in mind, diplomatic proposals might do well initially to focus on new ideas and rules that would be awkward for China to reject and would not turn on relative numbers per se.

It might be possible and useful, for example, to consider negotiations to establish a prohibition against using nuclear weapons against cities. This could make sense militarily and politically even if China did not agree.

First, China has more value and population concentrated in cities than the United States and so is arguably more vulnerable to attack than the United States. At the same time, China is more likely than the United States to threaten to target cities. Also, it would be useful diplomatically for the US to seek an agreement with China not to target cities in Japan and South Korea (two countries that are even more urbanized and vulnerable than China). Finally, pushing such an understanding would alert China's most urbanized irredentist provinces that in any war, the United States seeks to spare them. To the extent that the Chinese Communist Party leadership would be reluctant to risk losing these provinces in war, this message would indirectly strengthen America's hand in deterring the Chinese generally.

A second diplomatic initiative that might help Washington maintain alliance cohesion with Tokyo and Seoul and reduce the prospect of a massive Chinese nuclear ramp-up would be to have the United States, Japan, and South Korea approach Beijing about taking a time out on expanding "peaceful" uranium enrichment capacities and commercial spent fuel recycling plants. The logic of such an understanding (which could be the basis of a political deal) would be to trade Japanese, US, and South Korean nuclear fissile production restraint for similar Chinese caution.

This proposition should appeal to China, since Beijing is on record criticizing Japan for Tokyo's reprocessing plans (which include opening a massive plant at Rokkasho in 2022 capable of making 1,500 bombs' worth of nuclear explosive plutonium a year). Beijing would also prefer that South Korea, which has expressed an interest in building and operating large reprocessing and enrichment plants of its own, not to do so. Keeping both Japan and South Korea from developing such options is also in America's interest, as it would encourage both Asian allies to work more closely with Washington than they might otherwise.

Third, the United States and its Asian and European allies could start now to promote rules of the road for rendezvous satellite operations, ground-based lasers, and other anti-satellite jammers. These rules should be designed to reduce the potential use of these systems to launch surprise attacks against US and allied space-based assets critical to conducting nuclear and conventional military operations. These new rules also should clarify when, why, and how the right to self-defense might be exercised. The United States and its key NATO and East Asian allies would all have an interest in backing such rules to help clarify what actions might constitute acts of war in space. Commercial space firms (and commercial space insurance companies) internationally will also have an interest in adopting rules that would reduce unwanted space conjunctions, space debris, and uncertainties on the location of space objects generally.

As China becomes more dependent on space-based command, control, communication, surveillance and targeting systems, clarifying these matters should be of greater interest to Beijing as well. China wants to offer civil and commercial space services for export. Therefore, rejecting such rules might be awkward, as the rules would be useful not just to maintain safe military satellite operations, but to assure that commercial operations meet accepted international safety and insurance standards.

Finally, it might make sense to propose limiting the numbers of long-range hypersonic missiles on both sides. Since neither side is likely to deploy many of these weapons by 2030, China might have difficulty pushing aside such limitations. They also would allow the United States more time to develop (but not deploy) such weapons. After 2030, when the gap between China and the United States in deployed hypersonics is closed, Washington could safely agree on additional missile limits on shorter range hypersonics as well. By then, one would hope that the United States would have built a larger and more advanced missile production base than China's, one that could give America a margin of safety to build up if it had to.

Chapter 7:

Long-Term Prospects for Nuclear Missile Controls

David A. Cooper[487]

The long-term prospects for nuclear missile controls are as urgent as they are unpromising. Legacy international controls are moribund or unravelling. The paradigm of bilateral US-Russian nuclear arms control—which for decades has provided the world's only binding missile controls—is nearing the end of its rope. Meanwhile the global missile nonproliferation regime—always relatively weak—is increasingly ineffective in the face of steady missile technology diffusion and the mounting pressures of a dangerously fluid international security environment. Most alarmingly, these extant arrangements are woefully ill-equipped to control the emergence of new technologies, notably maneuverable hypersonic missiles.

The deterioration of international missile controls is nothing new. It has been happening in slow motion fits and starts for almost two decades. It is newly worrying, however, because the United States now confronts a new age of multipolar nuclear competition with a gathering missile arms race as its central feature. All the major nuclear powers are modernizing their nuclear missile forces, and most are racing to field revolutionary new types of missiles including intermediate- and intercontinental-range maneuverable hypersonic missiles, air launched ballistic missiles (ALBM), and in the case of Russia, a nuclear-armed and powered cruise missile of almost limitless

487. All views are solely those of the author and do not represent official positions of the US Naval War College, the Department of Navy, or any other agency or organization.

range. Now unshackled by the demise of the Intermediate Nuclear Forces (INF) Treaty, both Washington and Moscow are racing to deploy intermediate-range ballistic and cruise missiles—Russia having gotten a head start by blatantly violating INF—to offset China's overwhelming advantage in these dual-purpose conventional and nuclear systems. Meanwhile North Korea and Iran continue to expand and improve their already formidable arsenals of nuclear-cable missiles as they stretch toward intercontinental range. This volatile combination of missile racing among the major nuclear powers and nuclear and missile proliferation by hostile regional powers could provoke other countries to seek long-range missiles as a hedge for going nuclear later. In sum, the nuclear missile environment is deteriorating across the board with no end in sight.

This chapter explores the prospects for strengthening international controls on nuclear missiles over the next decade or more. The chapter begins by the exploring why missiles matter as organic components of nuclear forces and how this makes missile controls an essential feature of any viable nonproliferation or arms control regime. It then examines how the erosion of the post-Cold War order has opened the door for nuclear missile races in the face of increasingly contested regional and global nuclear landscapes. It then suggests the most plausible prospects for negotiating new or stronger nonproliferation and arms control measures to put guardrails on this increasingly uncontrolled global missile competition. Finally, it concludes by warning that these prospects are poor in today's fluid geopolitical and geostrategic circumstances and that the United States and its allies will need to play a long game to convince China, Russia, and others to come to the negotiating table.

Why Missiles Matter

When we talk about *nuclear weapons* we are really talking about missiles as well as the explosive warheads that they carry. Missiles are at the heart of both sides of the nuclear competition equation: new countries seeking to join the nuclear club (nuclear proliferation) and rivalry among the existing nuclear powers (nuclear postures and arms racing). Therefore, although the Nuclear Nonproliferation Treaty (NPT) does not restrict missiles, preventing nuclear proliferation nevertheless requires preventing missile proliferation. For countries like Iran and North Korea to become formidable nuclear powers they must not only obtain nuclear explosives, but also the means to deliver them. Likewise, nuclear balances among the major nuclear powers like Russia, the United States, and China are measured in large part by their missile delivery systems, and defenses against them, rather than warhead inventories *per se*. This is the case particularly for the United States and Russia, which have large reserve stockpiles of warheads and fissile materials to make more if necessary. To the extent that warheads are the focus, it is largely how many are operationally deployed on missiles. In a nutshell, nuclear delivery systems are a lynchpin of controlling nuclear dangers. President Ronald Reagan had it exactly right when he conflated things by referring to *nuclear missiles*.[488]

Ballistic and cruise missiles can have both conventional warfighting and nuclear deterrence roles. These roles tend to be defined by their ranges. Shorter-range systems are associated primarily with conventional warfighting and thus are seen as less relevant as nuclear delivery systems, although what counts as shorter-range is variously defined from below 300-1000 kilometers.[489] Medium-range systems (commonly defined as 1000-3000 kilometers) and intermediate-range systems (commonly defined as 3000-5500 kilometers) play both conventional and nuclear roles.[490] Although this may be changing, intercontinental missiles above 5500km to date have been exclusively nuclear missiles.[491] Despite these different uses, it is their

488. Ronald W. Reagan, Address to the Nation on Defense and National Security (March 23, 1983), https://www.reaganlibrary.gov/research/speeches/32383d.
489. For example, the Missile Technology Control Regime (MTCR) focuses on missiles over 300km, the Intermediate Nuclear Forces (INF) Treaty controlled missiles above 500km.
490. "Worldwide Ballistic Missile Inventories," Arms Control Association (December 2017), https://www.armscontrol.org/factsheets/missiles.
491. The US has considered using non-nuclear intercontinental ballistic missiles for Conventional Prompt Global Strike missions. However, it appears that a new generation of hypersonic boost-glide vehicles will fill this role. See: Amy F.

singular role as rapid and reliable nuclear delivery systems that sets ballistic and cruise (and now hypersonic) missiles apart from conventional weaponry, especially at longer ranges.

For this reason, the common distinction that is made between nuclear weapons and the missiles that carry them is not only misleading, but also misguided. The more inclusive concept of *nuclear forces* encompasses nuclear warheads and their delivery systems. For example, although the 1987 Intermediate Nuclear Forces (INF) Treaty was hailed as a breakthrough for nuclear disarmament, it was in fact a missile treaty that did not actually ban any nuclear weapons. Even long-range strategic bombers, like nuclear missile submarines, are really forward deployed staging platforms for long-range missiles. For example, the only reason that aging B-52s can still be effective against modern air defenses is down to the modern long-range air-launched cruise missiles (ALCMs) that they carry.[492] Indeed, other than comparatively antiquated nuclear gravity bombs intended for tactical nuclear warfighting, and Russia's new autonomous intercontinental-range nuclear torpedo (in effect an underwater missile), nuclear forces *are* missile forces. In other words, controlling nuclear forces mostly means controlling missiles.

Missile Proliferation Is Nuclear Proliferation

Despite the organic relationship between nuclear weapons and the missiles that carry them, missile nonproliferation is too often treated as separate and secondary to nuclear nonproliferation.[493] For exam-

---

Woolf, "Conventional Prompt Global Strike and Long-Range Ballistic Missiles: Background and Issues," Congressional Research Service (February 14, 2020), https://crsreports.congress.gov/product/pdf/R/R41464.

492. Aaron Mehta and Jeff Martin, "Decades Late, the B-52 is Getting a New Nuclear Weapon," *Defense News* (September 15, 2019), https://www.defense-news.com/smr/global-strike/2019/09/15/decades-late-the-b-52-is-getting-a-new-nuclear-weapon/.

493. This section draws significantly on a previous opinion piece and subsequent congressional testimony by the author and some portions are adopted directly from these earlier pieces: "The Real Iran Threat (Hint: It's Not Just Nukes)," *The*

ple, the nuclear deal that the Obama administration negotiated with Iran did not include any missile controls, leaving missiles to be addressed in a separate UN Security Council Resolution (UNSCR) that merely "called upon" Iran to suspend its missile activities—a request Iran has blithely ignored without violating the core agreement.[494] More significantly, the formal multilateral institutions responsible for preventing nuclear proliferation—the Nuclear Nonproliferation Treaty (NPT) setting the rules and the International Atomic Energy Agency (IAEA) verifying and implementing them—do not address missiles. This nuclear-only nonproliferation paradigm stands in sharp contrast to the Cold War nuclear arms control paradigm that focused holistically on weapons and delivery systems.

Focusing on long-range missiles brings into sharp focus that outright nuclear proliferation along the lines of North Korea and Pakistan is the tip of a potentially larger nuclear hedging iceberg—that is, countries that are deliberately acquiring the wherewithal to quickly become a breakout nuclear power should the need arise (sometimes also referred to as nuclear latency). Missiles play a key and integral role in nuclear hedging. It has long been understood that civilian nuclear energy programs can be used for nuclear hedging. For example, leading experts believe that Saudi Arabia and Japan are rather blatantly using their nuclear energy programs to hedge against growing nuclear threats from hostile neighbors.[495] But for these and

---

National Interest (March 23, 2015), https://nationalinterest.org/feature/the-real-iran-threat-hint-its-not-nukes-12461; "Iran's Enduring Missile Threat," testimony before the House Foreign Affairs Committee, (https://docs.house.gov/meetings/FA/FA13/20150610/103582/HHRG-114-FA13-Wstate-CooperD-20150610.pdf).

494. Resolution 2231 (July 20, 2015), p. 99, Section 3, https://undocs.org/S/RES/2231(2015).

495. James L. Schoff and Richard J. Samuels, "Japan's Nuclear Hedge: Beyond 'Allergy' and Breakout," in Abraham Denmark, Ashley J. Tellis, and Travis Tanner, eds., *Asia in the Second Nuclear Age* (Seattle: The National Bureau of Asian Research, 2013): 233-266; Sarah Burkhard, Erica Wenig, David Albright, and Andrea Stricker, "Saudi Arabia's Nuclear Ambitions and Proliferation Risks," Institute for Science and International Security (March 30, 2017), https://isis-online.org/uploads/isis-reports/documents/SaudiArabiaProliferationRisks_30Mar2017_Final.pdf; Toby Dalton, "Policy Implications of Nuclear Hedging:

other NPT countries there is still the safeguard (albeit imperfect) of IAEA oversight and the prohibition against moving beyond nuclear latency. By contrast, there is no international treaty that prohibits missile proliferation. Therefore, developing long-range missile capabilities—either overtly, or covertly within a civilian space launch program—is a much less restricted aspect of nuclear hedging.

Needless to say, there are some good reasons for prioritizing preventing the spread of nuclear weapons. First and foremost, even without missiles in the equation there is a strong case that nuclear weapons in the wrong hands pose a dire risk in that they could be used in a cataclysmic terrorist attack.[496] From the standpoint of purely military use, there is still always some risk posed by bombers or short-range missiles, especially for neighboring countries. Given that the expertise, equipment, and fissile materials that are required to produce nuclear weapons are extremely difficult to attain without outside assistance, thwarting nuclear weapons programs makes sense as a top nonproliferation priority. After all, if aspiring nuclear powers can be denied the wherewithal to produce nuclear weapons, then their aspirations will come to nothing with or without long-range missiles.

By the same token, however, a country that has nuclear explosives but lacks the capability to deliver them with missiles cannot be considered a formidable nuclear power. An underappreciate aspect of preventing nuclear proliferation is preventing the means for long-range and large-scale nuclear delivery. For instance, from a US perspective, while a nuclear-armed country without long-range missiles may menace American allies or overseas forces, it still does

---

Observations on East Asia," Carnegie Endowment for International Peace (October 10, 2019), https://carnegieendowment.org/2019/10/10/policy-implications-of-nuclear-hedging-observations-on-east-asia-pub-80169; Victor Gilinsky and Henry Sokolski, "To Prevent Proliferation, Stop Enrichment and Reprocessing in the Middle East," *Foreign Policy* (October 15, 2020), https://foreignpolicy.com/2020/10/15/to-prevent-proliferation-stop-enrichment-and-reprocessing-in-the-middle-east/;

496. Graham T. Allison, *Nuclear Terrorism: The Ultimate Preventable Catastrophe* (New York: St. Martin's Press, 2005).

not pose an existential threat to the American homeland. In other words, the threats posed by nuclear weapons *sans* long-range missiles boils down to local or terrorist threats. Giving short shrift to missile nonproliferation therefore makes little sense if the goal is to prevent countries like North Korea and Iran from becoming formidable entry-level nuclear powers like India and Pakistan.

How important are missiles for an aspiring nuclear power? Important enough that long-range missiles are a remarkably reliable indicator of any country's ultimate nuclear intentions. In fact, the lack of indigenous long-range missile programs is arguably the single most reliable indicator that a country does not harbor covert or latent nuclear ambitions—and vice versa. History shows that long-range missile programs tend to make military, political, and economic sense only in the broader context of an ambition to become a latent or actual nuclear power.[497] Nuclear weapons and long-range missiles go hand in glove, to the extent that no country has pursued nuclear weapons without also pursuing at least medium-range ballistic missiles. Conversely, countries that have reversed course on pursuing or possessing nuclear weapons, like South Africa and Libya, have also abandoned associated missile programs.[498]

It is difficult to overstate the correlation between long-range missiles—especially ballistic missiles—and nuclear weapons. No country without nuclear weapons or ambitions possesses intermediate-range ballistic missiles (IRBMs) or intercontinental ballistic missiles (ICBMS).[499] These longest-range systems are exclusively the domain of the existing nuclear powers and Iran.

---

497. An case in point is Argentina's abortive Condor II MRBM program. This was developed in the 1980s in part for military export sales, but was abandoned in the early 1990s due to international pressure. See declassified CIA memorandum: "Status of the Condor II Ballistic Missile Program" (November 1, 1991), https://www.cia.gov/library/readingroom/docs/DOC_0001175541.pdf.
498. Robert Joseph, *Countering WMD: The Libya Experience* (Fairfax, VA: National Institute Press, 2009); "South Africa: Missile," Nuclear Threat Initiative (April 2015), https://www.nti.org/learn/countries/south-africa/delivery-systems/.
499. "Worldwide Ballistic Missile Inventories," Arms Control Association (December 2017), https://www.armscontrol.org/factsheets/missiles.

As for medium-range systems, it is vanishingly rare for a country without nuclear weapons or ambitions to seek or possess missiles at this range. The few exceptions only serve to demonstrate the rule. Saudi Arabia is the only non-nuclear country with medium range ballistic missiles (MRBMs). However, these are not indigenous systems, but instead relatively antiquated missiles that Riyadh purchased decades ago 'off the shelf' from China.[500] Tellingly, the Saudis have long been suspected of nuclear hedging—with speculation that there are covert elements to the Kingdom's civilian nuclear energy program, or that Pakistan could be on tap to supply nuclear weapons for those missiles—and in recent years the Saudis now openly threaten to acquire nuclear weapons if Iran does so.[501] The other contemporary cases are South Korea and Taiwan, both of which are developing medium-range cruise missiles.[502] Yet these governments have unique in-depth warfighting needs against hostile nuclear-armed neighbors (and perhaps these new missile programs should raise questions about nascent nuclear hedging behavior). Past cases of countries pursuing longer-range missiles while proclaiming their non-nuclear status within the NPT include Iran and North Korea. In

---

500. "Saudi Arabia: Missile," Nuclear Threat Initiative (August 2015), https://www.nti.org/learn/countries/saudi-arabia/delivery-systems/.
501. Yoel Guzansky, "Saudi Arabia Nuclear Hedging," Atlantic Council (December 13, 2011), https://www.atlanticcouncil.org/blogs/new-atlanticist/saudi-arabia-nuclear-hedging/; Colin H. Kahl, Matthew Irvine, and Melissa Dalton, "Atomic Kingdom: If Iran Builds the Bomb, Will Saudi Arabia Be Next?" Center for New American Security (February 19, 2013), https://www.cnas.org/publications/reports/atomic-kingdom-if-iran-builds-the-bomb-will-saudi-arabia-be-next; Kingston Reif, "Saudi Arabia Threatens to Seek Nuclear Weapons," *Arms Control Today* (June 2018), https://www.armscontrol.org/act/2018-06/news/saudi-arabia-threatens-seek-nuclear-weapons; Andrea Stricker and Behnam Ben Taleblu, "Secret Sites in the Desert: The Dangers of Saudi Arabia's Nuclear Hedging," The National Interest (September 4, 2020), https://nationalinterest.org/feature/secret-sites-desert-dangers-saudi-arabia%E2%80%99s-nuclear-hedging-168267; Saudi Minister Says Nuclear Armament Against Iran 'An Option'," Aljazeera (November 17, 2020), https://www.aljazeera.com/news/2020/11/17/saudi-minister-wont-rule-out-nuclear-armament-over-iran.
502. CSIS Missile Defense Project: "Missiles of South Korea," https://missilethreat.csis.org/country/south-korea/; "Missiles of Taiwan," https://missilethreat.csis.org/country/taiwan/.

both cases the missiles turned out to be overt leading indicators of covert nuclear ambitions.

Missiles are not only integral to nuclear proliferation, but they also pose one of the most significant hurdles for wannabe nuclear powers to overcome. As North Korea and other cases demonstrate, it is often a faster feat to develop nuclear explosives than long-range missiles capable of reliably delivering them to distant targets. North Korea conducted successful nuclear tests before fielding what it claims are ICBMs that can hit the continental United States, and even those claims remain in serious doubt including whether these missiles can even carry a nuclear warhead.[503] This helps explain why, when Iran suspended its nuclear weapons program under the 2015 Joint Comprehensive Plan of Action (JCPOA), it eschewed missile constraints to continue to press forward on improving its missile capabilities.[504] Because building reliable and accurate long-range missiles that can carry nuclear weapons is as challenging as building those nuclear weapons. Indeed, one of the biggest *nuclear* challenges for new nuclear entrants like North Korea is designing nuclear warheads that are actually small and hardy enough to be delivered by ballistic missiles.[505] This ranking of difficulty tracks with the initial arc of nuclear history, in which the atomic age dates to the 1940s, the thermonuclear age to the 1950s, and the nuclear missile age to the 1960s.

As of now only the five recognized NPT nuclear powers have nuclear missiles at intercontinental ranges over 5500 kms, but others are

---

503. Choe San-Hun, "North Korea Unveils What Appears to Be New ICBM During Military Parade, *The New York Times* (November 27, 2020), https://www.nytimes.com/2020/10/10/world/asia/north-korea-icbm.html.
504. For an excellent analysis of the Iran missile challenge and potential remediation see: Robert Einhorn and Vann H. Van Diepen, "Constraining Iran's Missile Capabilities," The Brookings Institution (March 2019), https://www.brookings.edu/research/constraining-irans-missile-capabilities/.
505. Michelle Nichols, "North Korea has 'Probably' Developed Nuclear Devices to Fit Ballistic Missiles, U.N. Report Says," *Reuters* (August 3, 2020), https://www.reuters.com/article/us-northkorea-sanctions-un/north-korea-has-probably-developed-nuclear-devices-to-fit-ballistic-missiles-u-n-report-says-idUSKCN24Z2PO.

on a path to join this club.[506] India, Iran, North Korea, and Pakistan already deploy robust medium- and/or intermediate-range ballistic missile forces below the intercontinental threshold.[507] But deploying nuclear-armed ICBMs would put countries like Iran and North Korea in a different league in terms of being able to hold the American homeland at risk. In sum, the argument here is that nuclear and missile proliferation are effectively two sides of the same coin.

A Dangerous New Nuclear Missile Race

Preventing and countering nuclear proliferation by hostile regional powers has been the top priority of American nuclear policy for almost three decades. But now this focus is changing. In the coming decade worries about nuclear and missile proliferation are likely to be supplanted by a more existential global concern: the return of nuclear competition among the major nuclear powers. This is not to say that nuclear and missile proliferation are likely to recede in the years ahead, to the contrary. It is rather to say that despite the potential for regional proliferation spirals—particularly in the Greater Middle East and East Asia—these regional nuclear threats will be dwarfed by nuclear weapons' return to the center stage of international relations in a new multipolar era of great power competition. This is likely to be a qualitative technology race more than a numbers competition between the United States and Russia (although a narrower quantitative race could emerge for novel capabilities like long-range hypersonic missiles.) However, China and others such as

---

506. Sergio Pecanha and Keith Collins, "Only 5 Nations Can Hit Any Place on Earth With a Missile. For Now." *The New York Times* (February 7, 2018), https://www.nytimes.com/interactive/2018/02/07/world/asia/north-korea-missile-proliferation-range-intercontinental-iran-pakistan-india.html.
507. Defense Intelligence Ballistic Missile Analysis Committee, *2017 Ballistic and Cruise Missile Threat* (Wright-Patterson AFB, OH: National Air and Space Intelligence Center, June 2017); Kelsey Davenport, "Worldwide Ballistic Missile Inventories," Arms Control Association (December 2017), https://www.armscontrol.org/factsheets/missiles; CSIS, "Missiles of the World," CSIS Missile Defense Project, https://missilethreat.csis.org/missile/ (accessed September 26, 2020).

India and Pakistan are likely to continue racing to both modernize and expand and diversify their nuclear forces.[508]

Within American strategic circles it has been widely understood for half a decade or so that a new nuclear arms race is underway and gathering steam.[509] This is the first global nuclear arms race since the Cold War and only the second in nuclear history. However, it is utterly unprecedented in its character as the world's first tripolar nuclear arms race—between the United States, Russia, and China—in which *each* country is competing against *both* others. In fact, it can more properly be categorized as a multipolar arms race, because France and Britain are also competing with Russia (and potentially China) and India is also competing with China (as well as Pakistan). This unprecedented nuclear multipolarity is described by some as the dawn of a *third nuclear age*.[510] The realization that the world is facing a new nuclear arms race—and a perilous new *kind* of arms race—is even starting to seep into the popular imagination after decades of public inattention to nuclear dangers.[511] What is less ap-

508. *SIPRI Yearbook 2020: Armaments and International Security* (Oxford: Oxford University Press, 2020).
509. Author interviews in 2018-19 with nearly thirty former and current US officials. Also for an excellent overview of expert views on the current state of US-Russian arms racing see: Amy F. Woolf, "Russia's Nuclear Weapons: Doctrine, Forces, and Modernization," Congressional Research Service (July 2020), https://crsreports.congress.gov/product/pdf/R/R45861.
510. Patrick J. Garrity, "The Third Nuclear Age?" *Claremont Review of Books* (April 2015), https://www.claremont.org/download_pdf.php?file_name=2142Nuclear.pdf; Karl-Heinz Kamp, "Welcome to the Third Nuclear Age," *The National Interest* (May 2, 2016), https://nationalinterest.org/feature/welcome-the-third-nuclear-age-16020; Michal Smetana, "A Nuclear Posture Review for the Third Nuclear Age," *The Washington Quarterly* 41, no. 3 (Fall 2018): 137-157; Steven E. Miller, "A Nuclear World Transformed: The Rise of Multilateral Disorder, *Dædalus* 149 (special issue on "Meeting the Challenges of a New Nuclear Age"), no. 2 (Spring 2020): 17-36.
511. See for example: David Wroe, "Nuclear Club's Rival Weapons and Agendas Pushing Us Into Uncertain New World," *Sydney Morning Herald* (October 27, 2017), https://www.smh.com.au/politics/federal/nuclear-clubs-rival-weapons-and-agendas-pushing-us-into-uncertain-new-world-20171027-gz9jkq.html; Eric Schlosser, "The Growing Dangers of the New Nuclear-Arms Race," The New Yorker (May 24, 2018), https://www.newyorker.com/news/news-desk/

preciated, however, is that this new tripolar-cum-multipolar nuclear arms race mostly boils down to a missile technology race.

As in the Cold War nuclear arms race but even more so, today's calculations about adversarial nuclear balances—especially strategic forces capable of attacking a rival's homeland—center on quantity, quality, and capabilities (types) of nuclear delivery systems and defenses against them. To be sure, missiles have military uses other than nuclear. This has been increasingly the case for missiles with rangers over 500 kilometers that until recently were banned for the United States and Russia under the Intermediate Nuclear Forces (INF) Treaty. For example, China deploys thousands of such systems for area denial, most of which are thought to be for conventional warfighting.[512] For its part the United States has long been pursuing Conventional Prompt Global Strike (CPGS) capabilities—essentially intercontinental range conventional systems.[513] However, the race for new and better nuclear missiles lies at the heart of current long-range missile racing dynamics.

As of this writing, the New START Treaty appears set to be extended for a few more years as the last vestige of bilateral nuclear arms control. However, this brief and temporary reprieve will do little to alter the fundamental dynamics of a globe-spanning missile race that is for all intents already uncontrolled, New START notwithstanding. Not just Russia and the United States, but all the other existing

---

the-growing-dangers-of-the-new-nuclear-arms-race; William J. Perry and Tom Z. Collina, *The Button: The New Nuclear Arms Race and Presidential Power from Truman to Trump* (Dallas: BenBella Books, 2020); Andreas Kluth, "This Nuclear Arms Race Is Worse Than the Last One," Bloomberg (June 18, 2020), https://www.bloomberg.com/opinion/articles/2020-06-18/this-nuclear-arms-race-is-worse-than-the-last-one.

512. Robert Kobza, "Another Tool in the Toolbox: Using Intermediate-Range Missiles to Counter Chinese A2/AD in the Pacific," *Georgetown Security Studies Review* (December 2, 2019), https://georgetownsecuritystudiesreview.org/2019/12/02/another-tool-in-the-toolbox-using-intermediate-range-missiles-to-counter-chinese-a2-ad-in-the-pacific/.

513. James M. Acton, *Silver Bullet? Asking the Right Questions About Conventional Prompt Global Strike* (Washington, DC: Carnegie Endowment for International Peace, 2013), https://carnegieendowment.org/files/cpgs.pdf.

nuclear powers are moving aggressively to modernize their nuclear missile forces. Russia and China are also deploying new generations of road-mobile heavy intercontinental ballistic missiles (ICBMs) armed with multiple independent reentry vehicles (MIRVS) as well as exotic new missile variants such as air-launched ballistic missiles (ALBMs) and, in Moscow's case, a nuclear-powered and armed cruise missile capable of virtually unlimited range.[514] Moscow also claims that it has started to deploy its monster new Sarmat ICBM (also known as the Satan-2), which Russian state television boasts will be able to carry enough high-yield MIRVs for a single missile to obliterate the entire state of Texas or nation of France.[515]

Although Beijing is thought to have up to no more than a few hundred nuclear warheads to arm its much larger and mostly conventional missile forces, this estimate has long been contested by some prominent Russian analysts.[516] In any case, Beijing now fields a

---

514. Joseph Trevithick, "Russia Tests Modified RS-24 Ballistic Missile With an 'Experimental Warhead'," *The War Zone* (October 6, 2017), https://www.thedrive.com/the-war-zone/14941/russia-tests-modified-rs-24-ballistic-missile-with-an-experimental-warhead; Ankit Panda, "Revealed: China's Nuclear-Capable Air-Launched Ballistic Missile," *The Diplomat* (April 10, 2018), https://thediplomat.com/2018/04/revealed-chinas-nuclear-capable-air-launched-ballistic-missile/; Missile Defense Project, "DF-41 (Dong Feng-41 / CSS-X-20)," *Missile Threat* (August 12, 2016, last modified October 8, 2019), https://missilethreat.csis.org/missile/df-41/; Matthew Kroenig, Mark Massa, and Christian Trotti, "Russia's Exotic Nuclear Weapons and Implications for the United States and NATO," The Atlantic Council (March 2020), https://www.atlanticcouncil.org/wp-content/uploads/2020/07/Russias-Exotic-Nuclear-Weapons.pdf.

515. David Brennan, "Russia's 'Invulnerable' Satan 2 Nuclear Missile Will Be Ready to Fire by the End of 2020, Space Agency Official Says," *Newsweek* (July 8, 2019), https://www.newsweek.com/russia-satan-2-nuclear-missile-rs-28-sarmat-ready-fire-2020-1447994.

516. Victor Yesin, "China's Nuclear Capability," in Alexei Arbatov, Vladimir Dvorkin and Sergey Oznobishchev, eds., *Prospects of China's Participation in Nuclear Arms Limitations* (Moscow: Institute of World Economy and International Relations, 2012); Stockholm International Peace Research Institute, *SIPRI Yearbook 2019: Armaments, Disarmament, and International Security* (Stockholm: SIPRI, 2019); Office of the Secretary of Defense, Military and Security Developments Involving the People's Republic of China: Annual Report to Congress (September 2020), https://media.defense.gov/2020/

modern strategic nuclear triad and is on track to double its nuclear weapons arsenal over the next decade.[517] Moreover, senior US officials have indicated that China is driving toward nuclear parity with the United States.[518] China already enjoys an overwhelming advantage in intermediate-range ballistic and cruise missiles. Although these are mostly thought to be conventionally armed, many are dual-capable to carry either conventional or nuclear weapons.[519] This has Russia and the United States scrambling to develop and deploy new intermediate-range systems now that neither is constrained by the INF Treaty.[520]

Far more significant than modernizing and expanding existing types of long-range missile forces is the breakneck race among several major nuclear powers—China, Russia, the United States, France, and India—to develop and deploy intermediate- and intercontinental-range hypersonic missiles. Technically known as maneuverable hypersonic vehicles (MHVs), these systems can come in either

---

Sep/01/2002488689/-1/-1/1/2020-DOD-CHINA-MILITARY-POWER-REPORT-FINAL.PDF.

517. Office of the Secretary of Defense, Military and Security Developments Involving the Peoples Republic of China.

518. Ibid. (Office of the Secretary of Defense); "Transcript: Special Presidential Envoy Marshall Billingslea on the Future of Nuclear Arms Control," Hudson Institute (May 22, 2020), https://www.hudson.org/research/16062-transcript-special-presidential-envoy-marshall-billingslea-on-the-future-of-nuclear-arms-control; David Vergun, "China, Russia Nearing Status as US Nuclear Peers, Stratcom Commander Says," DOD News (July 30, 2020), https://www.defense.gov/Explore/News/Article/Article/2294574/china-russia-nearing-status-as-us-nuclear-peers-stratcom-commander-says.

519. Office of the Secretary of Defense, Military and Security Developments Involving the Peoples Republic of China.

520. Matthew Bodner, "Russia Bids Farewell to INF Treaty with Fresh Nuclear Development Plans," Defense News (February 6, 2019), https://www.defensenews.com/smr/nuclear-arsenal/2019/02/06/russia-bids-farewell-to-inf-treaty-with-fresh-nuclear-development-plans/; Kingston Reif and Shannon Bugo, "US Aims to Add INF-Range Missiles," *Arms Control Today* (October 2020), https://www.armscontrol.org/act/2020-10/news/us-aims-add-inf-range-missiles. Sydney J. Freedberg, Jr., "Army Seeks New Mid-Range Missile Prototype By 2023," *Breaking Defense* (September 8, 2020), https://breakingdefense.com/2020/09/army-seeks-new-mid-range-missile-prototype-by-2023/.

boost-glide (launched by missiles or planes and then gliding to target) or air-breathing cruise-missile varieties. MHVs combine the speed of a ballistic missile with the maneuverability of a cruise missile.[521] Hence, MHVs can reach targets faster and with less warning, and, even more important, they are virtually invulnerable to existing missile defense technologies.[522] Put bluntly, these systems promise to eclipse the capabilities of today's ballistic and cruise missiles. It is therefore no exaggeration that the coming hypersonic revolution puts global and regional nuclear balances completely up for grabs. Hypersonic missiles are therefore arguably the single most destabilizing aspect of the burgeoning new nuclear arms race.

It should be noted that only Russia has announced a strategic nuclear strike role for MHVs. In fact Moscow claims already to have started deploying what ultimately will be a force of sixty *Avangard* boost-glide MHVs on ICBMs, each armed with massive 2-megaton nuclear warheads.[523] Moscow also plans to begin arming its submarine-launched ballistic missiles (SLBMs) with nuclear MHVs by 2024, and may even deploy nuclear hypersonic missiles across its entire navy fleet.[524] Washington and Beijing also plan to deploy MHVs

521. It appears that some boost-glide MHVs will be able to fly faster than ICBMs. However, even ballistic missiles that can reach higher absolute speeds during the midcourse reentry phase of their flights will be slower to target than boost-glide MHVs due to their longer boost phase through the atmosphere.
522. Richard H. Speier, George Nacouzi, Carrie A. Lee, and Richard M. Moore. *Hypersonic Missile Nonproliferation Hindering the Spread of a New Class of Weapons* (Santa Monica: The RAND Corporation, 2017).
523. Russia Deploys Avangard Hypersonic Missile System," BBC (December 27, 2019), https://www.bbc.com/news/world-europe-50927648; Kyle Mizokami, "Russia's New Hypersonic Weapon Flies at Mach 27, Popular Mechanics (December 30, 2019), https://www.popularmechanics.com/military/weapons/a30346798/russia-new-hypersonic-weapon-mach-27/.
524. Amanda Macias, "Russian Submarine Fleet Capable of Launching Missiles Armed with Hypersonics and Nukes Will Be Ready for War by 2024," CNBC (September 21, 2018), https://www.cnbc.com/2018/09/21/russia-sub-fleet-capable-of-launching-hypersonics-will-be-ready-by-2024.html; "Andrew Osborn, Putin Says Russian Navy to Get Hypersonic Nuclear Strike Weapons," *US News and World Report* (July 26, 2020), https://www.usnews.com/news/world/articles/2020-07-26/putin-says-russian-navy-to-get-hypersonic-nuclear-strike-

at intermediate and intercontinental ranges.[525] Chinese hypersonic programs remain shrouded in secrecy and its intentions remain opaque in terms whether its MHVs will have a nuclear strike role. For its part Washington plans to deploy an array of MHVs starting in 2022-23 including road-mobile systems and MHVs launched from strategic bombers and fast-attack nuclear submarines.[526] Although it is unclear if any of these systems will be nuclear-armed there are indications that this possibility is at least being considered.[527] India and France round out the race for hypersonic missiles by nuclear powers, the latter moving quickly to deploy what is thought to be intended as new hypersonic nuclear delivery system.[528]

---

weapons.

525. John A. Tirpak, "Roper: Hypersonics Capability Less Than Two Years Away," *Air Force Magazine* (February 7, 2019); Andrew Tate, "China testing hypersonic weapon with intercontinental Range, Says USNORTHCOM Commander, Jane's Defense Weekly (February 18, 2020), https://www.janes.com/article/94373/china-testing-hypersonic-weapon-with-intercontinental-range-says-usnorthcom-commander; Kyle Mizokami, "Air Force Accidently Reveals It Wants a Hypersonic Nuke," *Popular Mechanics* (August 21, 2020), https://www.popularmechanics.com/military/weapons/a33657991/air-force-accidentally-reveals-interest-hypersonic-weapon-system/.

526. Colin Clark, "More B-21s Likely; B-1s to Carry Up To 8 Hypersonic Weapons," Breaking Defense (September 17, 2019), https://breakingdefense.com/2019/09/more-b-21s-likely-b-1s-to-carry-up-to-8-hypersonic-weapons/; Megan Eckstein, "Navy Confirms Global Strike Hypersonic Weapon Will First Deploy on Virginia Attack Subs," USNI News (February 18, 2020), https://news.usni.org/2020/02/18/navy-confirms-global-strike-hypersonic-weapon-will-first-deploy-on-virginia-attack-subs; Jen Judson, "How the DoD Plans to Meet its Ambitious Hypersonic Missile Test Schedule," *Defense News* (August 5, 2020), https://www.defensenews.com/digital-show-dailies/smd/2020/08/05/heres-how-the-dod-plans-to-meet-its-ambitious-hypersonic-missile-test-schedule/.

527. Kyle Mizokami, "Air Force Accidently Reveals It Wants a Hypersonic Nuke," *Popular Mechanics* (August 21, 2020), https://www.popularmechanics.com/military/weapons/a33657991/air-force-accidentally-reveals-interest-hypersonic-weapon-system/; Alan Cummings, "High Speed, Low Yield: A US Dual-Use Hypersonic Weapon," *War on the Rocks* (September 17, 2020), https://warontherocks.com/2020/09/high-speed-low-yield-a-u-s-dual-use-hypersonic-weapon/.

528. Michael Peck, "Now France Wants Hypersonic Missiles by 2021," *The National Interest* (February 4, 2019), https://nationalinterest.org/blog/buzz/

Meanwhile, today's nuclear missile race is not happening in splendid isolation. There is an accelerating military space race that threatens to give Russia and China antisatellite capabilities that could impact nuclear command and control..[529] Rapid developments in the cyber and artificial intelligence domains could also impact nuclear targeting, command and control, and crisis stability.[530] Meanwhile in a direct spinoff of the missile race, the threat of long-range hypersonic missiles is leading Washington to pursue space-based missile defenses against them.[531] But perhaps the most significant spinoff as this uncontrolled nuclear missile race accelerates could be to spark a new wave of nuclear missile proliferation. Controlling the nascent nuclear arms race and controlling nuclear proliferation thus may turn out to be intersecting goals.

There is every reason to suppose that a multipolar nuclear missile race could sooner or later provoke other great and/or regional powers to seek or expand their own missile arsenals as a nuclear hedge or even to breakout with nuclear weapons to arm them. This could include countries that currently rely on others to deter nuclear coer-

---

now-france-wants-hypersonic-missiles-2021-43202; H.I. Sutton, "India Goes Hypersonic: New Missile Technology May Be Answer To China's Navy," *Forbes* (September 8, 2020), https://www.forbes.com/sites/hisutton/2020/09/08/india-goes-hypersonic-new-missile-technology-may-be-answer-to-chinas-navy/#673ef44cd937.

529. Michael R. Gordon, "Russia Tests an Anti-Satellite Weapon, US Officials Say," *The Wall Street Journal* (July 23, 2020), https://www.wsj.com/articles/russia-tests-an-anti-satellite-weapon-u-s-officials-say-11595545670; Office of the Secretary of Defense, *Military and Security Developments Involving the Peoples Republic of China* (September 1, 2020), https://media.defense.gov/2020/Sep/01/2002488689/-1/-1/1/2020-DOD-CHINA-MILITARY-POWER-REPORT-FINAL.PDF.

530. Stephen J. Cimbala, "Nuclear Deterrence in Cyber-ia: Challenges and Controversies," *Air & Space Power Journal* 30:3 (Fall 2016): 54-62; James Johnson and Eleanor Krabill, "AI, Cyberspace, and Nuclear Weapons," *War on the Rocks* (January 31, 2020), https://warontherocks.com/2020/01/ai-cyberspace-and-nuclear-weapons/.

531. Kris Osborn, "The Pentagon Has Plans to Defend Against Hypersonic Missiles," *The National Interest* (October 10, 2020), https://nationalinterest.org/blog/buzz/pentagon-has-plans-defend-against-hypersonic-missiles-170543.

cion against them—e.g., allies under Washington's nuclear umbrella of extended deterrence—should they lose confidence in relying on others for nuclear security guarantees. A ramped up nuclear missile race could thus expand missile and nuclear proliferation beyond NPT scofflaws like Iran and North Korea to countries ranging from Australia, Germany, Japan, South Korea, Saudi Arabia, and Turkey.[532] For example Japan, which has long been seen to have a nuclear hedging strategy, is now developing an indigenous hypersonic missile.[533] It also has an extremely unusual three-stage, solid-fuel space launch vehicle that some experts believe gives Tokyo latent ICBM capabilities.[534] In other words, today's burgeoning nuclear missile race may herald the start of wider missile rivalries within and across key regions of geostrategic competition.

---

532. Matthew Karnitschnig, "German Bomb Debate Goes Nuclear," *Politico* (April 19, 2019), https://www.politico.eu/article/german-bomb-debate-goes-nuclear-nato-donald-trump-defense-spending/; Mark Fitzpatrick, "How Japan Could Go Nuclear," *Foreign Affairs* (October 3, 2019), https://www.foreignaffairs.com/articles/asia/2019-10-03/how-japan-could-go-nuclear; Byong-Chul Lee, "Don't Be Surprised When South Korea Wants Nuclear Weapons," *Bulletin of the Atomic Scientists* (October 23, 2019), https://thebulletin.org/2019/10/dont-be-surprised-when-south-korea-wants-nuclear-weapons/; Shannon Bugos, Turkey Shows Nuclear Weapons Interest," *Arms Control Today* (October 2019), https://www.armscontrol.org/act/2019-10/news/turkey-shows-nuclear-weapons-interest; Hugh White, *How to Defend Australia* (Carlton, Australia: La Trobe University Press, 2019); Emma Graham-Harrison, Stephanie Kirchgaessner and Julian Borger, "Revealed: Saudi Arabia May Have Enough Uranium Ore to Produce Nuclear Fuel," *The Guardian* (September 17, 2020), https://www.theguardian.com/world/2020/sep/17/revealed-saudi-arabia-may-have-enough-uranium-orc-to-produce-nuclear-fuel.
533. Mike Yeo, "Japan Unveils Its Hypersonic Weapons Plans," *Defense News* (March 13, 2020), https://www.defensenews.com/industry/tech-watch/2020/03/13/japan-unveils-its-hypersonic-weapons-plans/.
534. Saadia M. Pekkanen, "The Next-Generation Rockets that Japan Could Use to Protect Itself," *Forbes* (December 17, 2016), https://www.forbes.com/sites/saadiampekkanen/2016/12/17/the-next-generation-rockets-that-japan-could-use-to-protect-itself/?sh=5abac0dd40a3; Doug Tsuruoka, "Can Japan's Epsilon rocket be used as an ICBM?" *Asia Times* (23 January 2018), https://asiatimes.com/2018/01/can-japans-epsilon-rocket-used-icbm/.

The Need and Difficulties for New Controls

Preserving and strengthening the few existing guardrails on nuclear missile proliferation and arms racing faces daunting hurdles and negotiating new controls even more so due to broader and structural geopolitical and geostrategic trends. The systemic forces that are driving today's intensifying nuclear rivalries among the major nuclear powers—and which could exacerbate tomorrow's nuclear proliferation spirals—are part of a wider transition of the international system to a new era of great power competition. While not a unanimous consensus, disparate voices from across the ideological spectrum are converging on acceptance that the post-Cold War era of American global hegemony is over, or at the very least eroding beyond the point of no return.[535] But what new international order will replace the *Pax Americana* that has reigned since the Cold War ended three decades ago? This is the question that will shape the prospects and contours of nuclear competition and controls over the coming decade and beyond.

The question revolves mostly around the rise of China. The United States now believes that China is seeking to supplant it as the world's top economic and military power with malign designs on American interests and those of the wider liberal international order.[536] At the same time Russia is increasingly aligning with China geopolitically to work against the United States and its allies. Washington thus faces the challenge of competing against two rival great powers that are working together to undermine its global standing. Yet Moscow

---

535. Fareed Zakaria, "The Self Destruction of American Power: Washington Squandered the Unipolar Moment," *Foreign Affairs* 98, no. 4 (July/August 2019), https://www.foreignaffairs.com/articles/2019-06-11/self-destruction-american-power; Mathew J. Burrows, *Global Risks 2035 Update: Decline or New Renaissance?* (Washington, DC: Atlantic Council, 2019); Daniel Larison, "Please Tell The Establishment That US Hegemony is Over," *The American Conservative* (April 23, 2020), https://www.theamericanconservative.com/articles/somebody-tell-the-establishment-that-american-hegemony-is-over/.
536. Policy Planning Staff, "Elements of the China Challenge," Office of the Secretary of State (November 2020), https://www.state.gov/wp-content/uploads/2020/11/20-02832-Elements-of-China-Challenge-508.pdf.

and Beijing still have latent geostrategic tensions with each other. Some experts argue that superficial Sino-Russian cooperation masks deeper fissures that will erode this geopolitical marriage of convenience over time.[537] There is therefore still a tripolar nuclear rivalry underlying a nascent bipolar alignment that may or may not endure through the swirl of broader global power realignments.

The difficulty for analysts and statesmen alike is that almost everything about the rise of China and its potential geostrategic reverberations remains up in the air. Will China rise to surpass the United States as the world's leading power as many assume or is it already showing signs of faltering as some contrarians suggest?[538] If China does become the leading global power, then could its ambitions be checked by other powers that bandwagon with the United States to balance against it? Or will a bipolar order emerge akin to the Cold War with China and the United States leading opposing ideological coalitions? Or might all this turn into a more fluid multipolar system akin to the 19th Century's Concert of Europe as various powers bandwagon with or balance against China and the United States to preserve a balance of power? Within these scenarios will Russia

---

537. Jack Farchy, "China's Great Game: In Russia's Backyard," *Financial Times* (October 14, 2015); Michal Makocki and Nicu Popescu, *China and Russia: An Eastern Partnership in the Making?* (Chaillot Paper no. 140) (Paris: European Union Institute for Security Studies, December 2016), www.iss.europa.eu/sites/default/files/EUISSFiles/CP_140_Russia_China.pdf; Andrew McCormick, "Russia is Using Its Military to Become a Major Player in China's Backyard, and it Could Mean Trouble if Moscow Pushes Too Far," *South China Morning Post* (September 5, 2018); Leon Aron, "Are Russia and China Really Forming An Alliance? The Evidence is Less Than Impressive," *Foreign Affairs* (April 4, 2019), www.foreighaffairs.com/articles/china/2019-04-04/are-russia-and-china-really-forming-alliance; Elizabeth Buchanan, "There's No (New) China–Russia Alliance," *The Strategist* (June 26, 2020), https://www.aspistrategist.org.au/theres-no-new-china-russia-alliance/.
538. Michael Beckley, "The United States Should Fear a Faltering China: Beijing's Assertiveness Betrays Its Desperation," *Foreign Affairs* (October 28, 2019), https://www.foreignaffairs.com/articles/china/2019-10-28/united-states-should-fear-faltering-china; Andrew Latham, "Has China Peaked?" *The National Interest* (November 19, 2020), https://nationalinterest.org/feature/has-china-peaked-172911.

continue to be drawn further into the Chinese orbit or will historic tensions reemerge between the two land giants of Asia? Will Europe finally coalesce as a unified superpower to rival the United States and China or will it fracture into less significant geostrategic parts? Will India rise to become a major axis of global influence as a first-tier great power? As all of this plays out will America's allies feel confident continuing to rely on Washington for their security or will they increasingly feel the need to fend more for themselves?

Because such basic questions about the emerging world order are unanswerable with any high degree of certainty, hedging behavior and security dilemmas are likely to abound among the major global and regional powers until a clearer picture comes into focus. Given that most of the primary players in this great power jockeying other than Japan are already major nuclear powers—America, China, Russia, Europe (Britain and France) and India—this is not a good recipe for a preserving a stable nuclear landscape; all the more so because today's nuclear multipolarity is asymmetric and misaligned relative to broader measures of global power. Russia is the poorest and militarily weakest of the leading great powers but also has the world's largest and most formidable nuclear arsenal. This includes a massive advantage in tactical systems designed for nuclear warfighting rather than strategic deterrence. The United States still has a large strategic nuclear force but has fallen far behind Russia and China on strategic modernization. China is rising to become the world's leading power but still is presumed to have much smaller nuclear forces than either Washington or Moscow even as it presses forward with aggressive nuclear expansion. Again, these asymmetries do not augur well for nuclear cooperation or stability.

The upshot is that none of the major nuclear powers, or nuclear aspirants, or countries that may prefer to leave the door open to acquiring long-range missiles for nuclear hedging purposes, have common or even similar incentives to cooperate on new missile controls. This makes any multilateral missile nonproliferation treaty about as far-fetched as a sweeping disarmament agreement among the major nuclear powers to reduce or eliminate their nuclear arsenals. The most realistic approach in these unpromising circumstances is to focus on preserving what can be preserved and seeking modest and/or narrow new measures to address the most urgent gathering perils. This may require bringing sustained international pressure to bear on reluctant nuclear powers like China—which has a mixed record in supporting nuclear nonproliferation and rejects participating in nuclear arms control out of hand—and other countries pursuing long-range missile programs. The most optimistic long-term goals would be to strengthen the existing missile nonproliferation regime while in parallel reinventing the bilateral Cold War nuclear arms control paradigm to fit today's asymmetric tripolar-cum-multipolar nuclear arms race.

Strengthening Missile Nonproliferation

The international missile nonproliferation regime primarily comprises a pair of rather weak arrangements, one an informal supply-side mechanism to coordinate export controls and the other a feeble global norm against ballistic missile proliferation. The supply-side of this equation is a loose arrangement among thirty-five partner countries called the Missile Technology Control Regime (MTCR). The demand-side is a broad multilateral agreement open to all countries called the Hague Code of Conduct (HCOC) Against Ballistic Missile Proliferation (sometimes also known as the International Code of Conduct or ICOC). While neither of these arrangements are especially robust, the MTCR is by far the more effective of the two.[539] The question is whether these existing missile nonproliferation pillars can be strengthened and if there are workable prospects for any new controls to augment them.

There is certainly some scope to strengthen the MTCR, although probably only on the margins. By contrast HCOC is likely too thin a reed to build upon as such. That said, it could provide a useful

539. For an excellent discussion of their relative strengths and weaknesses see: Andrew Feickert, "Missile Technology Control Regime (MTCR) and International Code of Conduct Against Ballistic Missile Proliferation (ICOC): Background and Issues for Congress," Congressional Research Service (April 8, 2003), https://crsreports.congress.gov/product/pdf/RL/RL31848/3.

starting point as a venue in which to elevate the profile of missile nonproliferation. The best opportunity to bolster nuclear missile nonproliferation, however, may be to try to try to negotiate a new, narrowly focused, global nonproliferation regime to nip the longest-range hypersonic missiles in the bud. Finally, the looming proliferation challenges suggest the need to look beyond diplomatic solutions to bolster extended deterrence by the United States (and perhaps others) to provide a disincentive for certain non-nuclear great and regional powers such as Japan, South Korea, and Germany to pursue long-range missiles as a nuclear hedge. These are limited and imperfect options, but they are worth exploring.

Strengthening the MTCR is low hanging fruit. This export control arrangement arose from a US initiative with six of its close allies in the 1980s to coordinate national export controls on missiles and related dual-use technology.[540] The regime comprises a set of agreed guidelines that provide for a "strong presumption of denial" on exports of any missile that can carry a 300kg payload to ranges above 500km and vigilance on exporting periodically updated lists of dual-use materials and technologies that could contribute to such missiles.[541] The MTCR has worked reasonably well as far as it goes, but its long-term potential suffers from partners that are not always on the same political page. This is not surprising considering that it has expanded over the years to include countries that often do not see eye to eye with the United States and its allies on various issues, such as Russia, South Africa, and Brazil. Because the MTCR operates by strict consensus, any of these countries can block any proposal intended to tailor, expand, or strengthen the scope of con-

---

540. For details on the early history of the MTCR see Ibid. and: David A. Cooper, "The United States and the Evolution of Supply-Side Missile Non-Proliferation Controls," in *Missile Proliferation and Defense: Problems and Prospects* (Monterey, CA: James Martin Center for Nonproliferation Studies, 2001): 15-20, https://www.jstor.org/stable/resrep09897.8?seq=1#metadata_info_tab_contents.
541. MTCR Guidelines for Sensitive Missile-Relevant Transfers, https://mtcr.info/guidelines-for-sensitive-missile-relevant-transfers/; MTCR Equipment, Software and Technology Annex (October 11, 2019), https://mtcr.info/wordpress/wp-content/uploads/2019/10/MTCR-TEM-Technical_Annex_2019-10-11-1.pdf.

trols. Moreover, enforcement relies on good faith interpretation and implementation by each partner country, which naturally invites potential differences. Even the United States itself has been reluctant to enforce some MTCR rules, although Washington argues that this is because inertia has prevented these from being updated to reflect current technological and national security realities.[542] In effect the MTCR is a useful technical regime that is often stymied by political inertia and the looseness of its structure and rules.

One obvious workaround for some of these issues would be to retain the MTCR while in parallel the United States seeks to organize a smaller caucus of its closest and most important allies to see if there is scope for this core group to go further amongst themselves than the more unwieldy larger group is willing or able. This would amount to creating a new regime within a regime. Such an *informal core group* could meet initially at higher political levels to scope out areas for enhanced cooperation. These could include going beyond what is strictly required by the MTCR rules in order to impose stricter controls on certain missiles, for example stealthy, hypersonic, or longer-range systems. Core group members could implement these among themselves while working as a unified caucus to encourage to wider group to follow suit. A core group could also agree among themselves to subject countries of particular proliferation concern like Iran to additional scrutiny. They could also agree to enhanced implementation of the existing MTCR rules by subjecting especially sensitive export licenses to *post-licensing verification* through export license terms that require recipients to allow on-site inspections to guard against items being diverted to improper uses. This is an underutilized verification technique that currently only the United States implements in any significant way. None of these are new ideas.[543] Nor are they especially dramatic. Instead, they are

---

542. Daryl G. Kimball, "US Reinterprets MTCR Rules," *Arms Control Today* (September 2020), https://www.armscontrol.org/act/2020-09/news/us-reinterprets-mtcr-rules.
543. David A. Cooper, "Hedging the Iran Nuclear Bet: Reinvigorate Supply-Side Nonproliferation," *The Washington Quarterly* 39, no. 3 (2016): 41-59, https://www.tandfonline.com/doi/abs/10.1080/0163660X.2016.1232633?journalCode=rwaq20.

modest technical and process improvements. But that is the nature of multilateral export controls— the unglamorous workhorses of supply-side nonproliferation. It is also as much as the political traffic of thirty-five disparate partner countries is likely to bear. These incremental supply-side improvements could nonetheless be useful to impede the further diffusion of missile technology and in any case are more likely to bear fruit than anything on the normative side of the equation.

There is no global missile nonproliferation treaty. Nor is there likely to be one for the foreseeable future given the reliance of the NPT nuclear weapons states (and nuclear powers outside NPT) on long-range missiles, in conjunction with the reluctance of many other countries to agree to another discriminatory treaty along the lines of the NPT that provides for haves and have-nots. The closest thing to a demand-side global norm is the HCOC, but it is an utterly feckless arrangement that has no enforcement mechanisms and, in any case, does not actually prohibit much of anything and does not even address cruise or hypersonic missiles.[544] It is at most a normative confidence building measure to demonstrate hortatory support for ballistic missile nonproliferation. For all of these conspicuous shortcomings though, it could still be used as a platform to stir international attention on missile proliferation. What is offers in this regard are annual meetings of a majority of the world's countries pledging to work together against ballistic missile proliferation. Although HCOC now boasts 140 subscribing states, there are key countries that are conspicuous by their absence, including China, Egypt, India, Iran, Israel, Saudi Arabia, and Pakistan.[545] *Elevating the level of HCOC meetings* could provide opportunities to advance ideas that would bring helpful political pressure to bear on these holdouts. Annual HCOC meetings have always been obscure bureaucratic af-

544. United Nations General Assembly Document A/57/724 (February 6, 2003), https://us02web.zoom.us/j/85384886824?pwd=VjRxMjBKRVQrUHBGazliTm16SThIdz09.

545. Bureau of International Security and Nonproliferation, "The Hague Code of Conduct Against Ballistic Missile Proliferation," US Department of State (April 2019), https://www.state.gov/hague-code-of-conduct-against-ballistic-missile-proliferation-hcoc/.

fairs, but this need not be the case. The United States could easily up its representation in coordination with other key players to introduce ambitious proposals that align with the goals and purposes of the agreement.

One such initiative might be to work with Moscow to use HCOC as a vehicle to revive a 2007 joint US-Russian proposal to *globalize the now defunct INF Treaty* that called on all countries to join them in "renunciation of ground-launched ballistic and cruise missiles with ranges between 500 and 5,500 kilometers, leading to the destruction of any such missiles, and the cessation of associated programs."[546] The United States and Russia in effect would offer to rejoin the INF Treaty—with its basic terms non-negotiable—but only if others follow suit. To be clear this proposal would stand no chance of gaining traction given that China enjoys a huge advantage in these types of systems and that they represent the mainstay of the missile forces of countries like Iran, North Korea, and India. However, an actual treaty could be negotiated with willing countries by incorporating a provision into the relevant INF prohibitions that would require all countries with prohibited missiles or programs to join before the treaty enters into force, along the lines of a similar provision in the Comprehensive Test Ban Treaty (CTBT). This would allow the United States to adopt a 'we'll sign if you do' stance with those already deploying systems that would be banned.[547] This would set the stage for a long-term norm-building effort. By offering to reconstitute INF as a global missile nonproliferation and disarmament treaty, such an initiative could provide a rare point of alignment between Washington and Moscow against Beijing. It would afford Washington the

546. Office of the Spokesman, "Joint US-Russian Statement on the Treaty on the Elimination of Intermediate-Range and Shorter-Range Missiles at the 62nd Session of the UN General Assembly (October 25, 2007)," US Department of State Archive, http://2001-2009.state.gov/r/pa/prs/ps/2007/oct/94141.htm.

547. Washington might even opt to go so far as to sign but not ratify as it has done with CTBT. The complication here is that a treaty signatory is bound to refrain from actions that would undermine the object and purpose of that treaty. In order to sign (or sign and ratify), the treaty would need to be negotiated to explicitly permit intermediate-range missile deployments pending its entry into force.

nonproliferation high ground while shining the spotlight on China and others as the root problem, even as Washington and Moscow reconstitute their own post-INF intermediate missile forces. This is a longstanding idea that has never gone anywhere and there is a good chance that even Russia might now shy away for fear of straining its cozy relations with Beijing. Then again, Russia has more reason to be concerned about these Chinese missiles—which can hit it territory but not the United States—and this was initially a high-profile proposal by President Vladimir Putin that he might find difficult to disavow.[548]

Another nonproliferation initiative urgently worth considering is seeking to negotiate a *global ban on long-range and nuclear-armed MHVs.* At this point only a few countries have long-range hypersonic missile programs in advanced stages of development and all of them other than Japan are existing nuclear powers. Because the most feasible nonproliferation measures are prophylactic this should be an urgent priority before other countries join the hypersonic race. This technology is likely to prove too critical for conventional warfighting—like cruise missiles in times past—to make a complete ban practicable. However, there is no more reason for non-nuclear states to have long-range MHVs than there is for them to have long-range ballistic missiles. This would be simple to verify for states that do not have MHVs in the form of a test ban. The real challenge is that the nuclear powers may insist on retaining intermediate- and intercontinental MHVs as nuclear delivery systems. If so, this would necessitate a discriminatory 'NPT-like' approach that exempts the NPT nuclear weapons states and/or countries that are already in advanced stages of developing such systems. This would be an extremely tough sell to everyone else. It would also necessitate leaving out the non-NPT nuclear powers or else conferring de facto legitimacy on their nuclear status (always assuming that the NPT itself does not unravel if states harboring nuclear ambitions ever threaten or opt to withdraw). Perhaps the best way to navigate this would be to follow the NPT model of exempting only countries that have al-

548. David A. Cooper, "Globalizing Reagan's INF Treaty: Easier Done Than Said?" *The Nonproliferation Review* 20, no. 1 (2013): 145-163.

ready tested long-range MHVs. In any case, this scheme could only work if the pot could be sweetened for prospective have-nots by having exempted countries undertake a parallel arms control measure to cap their deployments of long-range and/or nuclear-armed MHVs (more on this idea below).

Finally, the dubious prospects for any of these initiatives suggests the importance to pursue another means to strengthen missile nonproliferation that lies beyond the realm of diplomacy: *repairing and expanding extended deterrence*. As noted already, a serious long-term risk for missile proliferation might come not only from the usual rogue suspects but also American allies. If these countries lose confidence in Washington's security guarantees, then they might pursue long-range missiles as part of a nuclear hedging or breakout strategy. Extended deterrence has always been the secret sauce of nuclear nonproliferation by reassuring most American allies that they do not need their own nuclear forces. Unfortunately, faith in the US nuclear umbrella has been undermined over the past decade; first by the Obama administration's perceived anti-nuclear stance including public flirtations with 'sole use' and 'no first use' doctrines, then by the Trump administration's perceived anti-alliance stance including flirtations with withdrawing or disavowing the automaticity of promises to respond to attacks against allies (e.g., NATO Article 5).

It is left to the Biden administration and its successors to repair this damage. But this may be challenging if President Biden seeks to revisit 'sole use' or 'no first use' declaratory policies, as he will doubtless be prevailed upon to do by dovish factions within his party. As noted, extended deterrence has been buffeted from different directions by the last two administrations. What is therefore needed is to couple rebuilding allied confidence in American support for NATO and its other key alliances (as President Biden has pledged) with not only a strongly reaffirmation of existing nuclear guarantees, but even expanding these judiciously through steps like declaring Australia unambiguously under the American nuclear umbrella. There could even be useful scope for other nuclear powers to use extended deterrence to calm skittish allies. There are already suggestions that France should consider extending its own independent nuclear guarantees to the rest of Europe to compensate for worries about

NATO, and President Emmanuel Macron has articulated a tough-minded nuclear posture that does not seem to rule out this possibility.[549] Would it even be so crazy to think that formal Chinese nuclear guarantees to Pyongyang might be part of a solution to the elusive North Korean denuclearization puzzle?

Reinventing Nuclear Arms Control

The odds for major new arms control agreements in the context of competitive and uncertain nuclear multipolarity are negligible in the near-term and even more distant prospects are not that much more encouraging. However, there are some modest interim options and aspirational long-term proposals that might help over time to stabilize deterrence relationships amid the gathering nuclear arms race. If the New START Treaty gets extended there will be a measure of transparency and predictability for at least a few more years in the adversarial nuclear relationship between the United States and Russia. This brief interval should not be squandered. It should be used to probe the scope to reinvent the longstanding model of bilateral nuclear arms control for the new circumstances of competitive nuclear multipolarity. It means setting aside any dreams of nuclear abolition, or even steep reductions, to focus on more modest measures to calm arms racing and stabilize strategic deterrence relationships. It also means that the future of nuclear arms control—which is mostly nuclear missile control—must be trilateral or plurilateral. This will require President Biden embrace and build on the legacy of his predecessor to draw China into trilateral negotiations, perhaps a bit-

549. Christian Trippe, "Could France Take the Lead in Europe's Nuclear Security?" *DW* (February 16, 2019), https://www.dw.com/en/could-france-take-the-lead-in-europes-nuclear-security/a-47549878; Emmanuel Macron, Speech of the President of the Republic on the Defense and Deterrence Strategy (February 7, 2020), https://www.elysee.fr/en/emmanuel-macron/2020/02/07/speech-of-the-president-of-the-republic-on-the-defense-and-deterrence-strategy; Ivon Dikov, "Macron Just Missed the Chance to Extend France's Nuclear Umbrella over the Post-Brexit EU," *European Views* (February 27, 2020), https://www.european-views.com/2020/02/macron-just-missed-the-chance-to-extend-frances-nuclear-umbrella-over-the-post-brexit-eu/.

ter political pill to swallow. However, arms control among nuclear rivals will only work if it aligns with geopolitical and geostrategic realities. This is precisely why the traditional bilateral Cold War approach withered on the vine in the unipolar post-Cold War order—it did not align with this unipolar reality. In a multipolar nuclear order, the future of arms control can only be trilateral or plurilateral.

The minimum requirement for any future nuclear arms control process is to at least cover the tripolar Sino-Russo-American core of global great power competition. It may ultimately need to expand to plurilateral arrangements if Russia continues to insist that Britain and France participate, and/or if China adopts a similar stance about India. Regardless of whether the format is trilateral or plurilateral, the salient point is that New START represents the denouement of bilateral strategic arms control. China's argument that its smaller nuclear arsenal exempts it from arms control will not wash over time. In the first place there is uncertainty about the precise size of the Chinese arsenal other than that it is growing. More important, China is now the main geopolitical and geostrategic rival of the United States. If new arms control is to happen then China must be in the mix. President Obama made this point as far back as 2009 by indicating that the purpose of the then ongoing New START negotiations was to prime the pump for expanding nuclear negotiations to the other nuclear powers.[550] Extending New START must be seen in this same light as when it was first negotiated.

China is not going to come willingly to arms control. It adamantly rebuffed the Trump administration's efforts to draw it into a trilateral negotiating process.[551] Beijing therefore needs to be pressured over time to see that participating in arms control is in its best interests. At the same time Russia will continue to be opportunistic about its

550. Barack Obama, "Remarks By President Barack Obama In Prague As Delivered," Office of the Press Secretary (April 5, 2009), https://obamawhitehouse.archives.gov/the-press-office/remarks-president-barack-obama-prague-delivered.
551. "Envoy: China Won't Take Part in US-led Trilateral Arms Control Negotiation," *CGTN* (October 14, 2020), https://newsus.cgtn.com/news/2020-10-14/Envoy-China-won-t-take-part-in-negotiating-trilateral-arms-control-Uz1nRZo2Hu/index.html.

willingness to engage in new arms control and so will also need to be incentivized. This situation requires the United States to arms race against China and Russia while holding out the offer to negotiate a trilateral or plurilateral arms control alternative. Why? Because Russia and China are only going to engage in good faith arms control negotiations if and when they believe that the only alternative is a dangerous and expensive arms race that they stand to lose (or at least that they cannot realistically hope to win).

Unfortunately, the United States is not able to negotiate from strength at the moment. Washington is behind Moscow numerically in strategic deployments (due to some disadvantageous quirks in New START's counting rules) and far behind in total nuclear deployments (due to Moscow's massive advantage in nonstrategic systems). It is likewise far behind both Russia and China in nuclear modernization. Moreover, unless the United States successfully modernizes its entire nuclear force within the next 10-15 years, senior Pentagon officials warn that the United States could cease altogether to be a credible nuclear power.[552] This lagging position is compounded by partisan political divisions that might encourage America's rivals to take a wait and see approach to arms control. Only when Washington demonstrates a sustained bipartisan commitment to nuclear modernization through one or more election cycles will we be able to see what new nuclear guardrails may be possible. The initial stance of the Biden administration will be crucial in this regard.

The Biden administration would do well to propose aspirational arms control initiatives that are nevertheless more realistic than the deep cuts and nuclear abolition goals previously pursued by the Obama administration. In parallel, Washington must continue to demonstrate that it is willing and able to: 1) continue fully to fund and execute the nuclear modernization program of record; 2) match and surpass Russian and Chinese long-range hypersonic systems absent

552. C. Todd Lopez, "'Tipping Point' Is Here for Nuclear Modernization, Defense Official Says," DOD News (September 17, 2020), https://www.defense.gov/Explore/News/Article/Article/2351959/tipping-point-is-here-for-nuclear-modernization-defense-official-says/.

new control measures; and 3) press forward with nascent plans for a space-based missile defense sensor architecture to defend against them (again, absent new control measures). Chinese resistance to intrusive verification suggests that any new initiatives should focus squarely on delivery systems, which are easier to verify than warheads. It is instructive to recall that SALT-I—lacking intrusive verification measures—mainly controlled only launchers (e.g., ICBM silos and SLBM launch tubes). What follows are some plausible suggestions for specific arms control proposals along these lines.

Washington could propose a *US-Russian intermediate-range deployment moratorium in Europe.*[553] This geographic deployment ban would prohibit range ground-launched missiles of all types (ballistic, cruise, and hypersonic) with ranges from 500-5500 kilometers from the Atlantic to the Urals. This would remove theater missiles from Western Europe and Russia's European territory. The goal would be to avoid a missile race in Europe by shifting Russian missiles and any potential US overseas deployments toward Asia. This could be achievable because, if NATO opts to deploy intermediate-range missiles to counter Russian deployments, Russia will find itself at a disadvantage relative to the United States. Why? Because these missiles will be able to hit important targets in the Russian homeland whereas Russian missiles could not target the US homeland. To move quickly and to minimize the risks posed by any potential Russian violations this could be proposed as reciprocal unilateral moratoria rather than as a formal bilateral treaty. If Russia agrees, then this would reassure NATO allies while ramping up pressure on China to negotiate trilaterally. If Russia refuses, then Washington will have lost nothing and gained political points for trying. It may or may not work, but there are really no downsides.

The Biden administration should also consider moving quickly and urgently to propose an *outright ban or low ceilings for strategic-*

553. This proposal draws on a prior opinion piece by the author: "Could a U.S.-Russian Deployment Ban Avoid a Nuclear Arms Race in Europe?" *The National Interest* (April 2, 2019), https://nationalinterest.org/feature/could-us-russian-deployment-ban-avoid-nuclear-arms-race-europe-50457.

range and/or nuclear-armed MHVs. As noted earlier, preemptive controls are typically the most negotiable. It is too late to preemptively negotiate a total ban on all types and ranges of MHVs given that several nuclear powers have made great strides already in developing such systems including China, Russia, the United States, France, and India.[554] MHVs are also too potentially useful for conventional strike missions for a total ban to be workable, especially at shorter ranges. However, a narrower ban on strategic-range (either intermediate or intercontinental) and/or nuclear-armed MHVs would not seem to be out of the question. Russia might balk at a total ban given that it claims already to have started to deploy such systems. The United States may also be reluctant to give up on plans to use MHVs for conventional global strike. China may hope that it can cling to the leading edge in developing new hypersonics. But none of these programs are far enough along to rule out a narrow ban. If a total ban proves out of reach, then national ceilings for strategic-range and/or nuclear-armed MHVs could still enhance deterrence stability, especially if these were negotiated below the level that any country thought would put its nuclear deterrent it at risk from a first strike. Low enough ceilings might also convince the rest of the world to renounce such systems for purposes of missile nonproliferation as discussed above.

In addition to overall long-range MHV ceilings there could be *ceilings on multiple reentry vehicles* encompassing both multi-MHV boosters and traditional MIRVs. This could be achieved by an overall ceiling on deployed warhead numbers relative to the number of strategic delivery systems. Alternatives might be to cap the maxi-

---

554. Michael Peck, "Now France Wants Hypersonic Missiles by 2021," *The National Interest* (February 4, 2019), https://nationalinterest.org/blog/buzz/now-france-wants-hypersonic-missiles-2021-43202; Richard Stone, "'National Pride is at Stake.' Russia, China, United States Race to Build Hypersonic Weapons," *Science Magazine* (January 8, 2020), https://www.sciencemag.org/news/2020/01/national-pride-stake-russia-china-united-states-race-build-hypersonic-weapons; Sarah Zheng, "India Joins Hypersonic Club with Successful Test Flight of Cruise Vehicle," *South China Morning Post* (September 8, 2020), https://www.scmp.com/news/china/diplomacy/article/3100662/india-joins-hypersonic-club-successful-test-flight-cruise.

mum number of MIRVs or multi-MHVs that can be deployed on any single ICBM and/or the maximum number of launchers that can hold MIRVed or multi-MHV missiles. This would temper current trends pointing toward a destabilizing surge in MIRV and multi-MHV systems. Russia is re-MIRVing aggressively and plans multi-MHV ICBMS. China and India are now MIRVing for the first time. The United States currently MIRVs only SLBMs and thus is not yet in this race. But it could gain significant negotiating leverage just by pointing to its ability quickly to re-MIRV should it choose to do so by uploading stored warheads back onto deployed ICBMs. Realistically, Russia is unlikely to forgo its MIRV and nascent multi-MHV advantages. and China is unlikely to countenance the intrusive measures that verifying any MIRV ceiling would necessitate. Still, it would be worthwhile to put this out there as an aspirational proposal.

Conclusion

Nuclear missiles and defenses against them are central to the US-Russia-China strategic rivalry that is gaining momentum and pulling in the secondary nuclear powers. In particular, the technology race for hypersonic missiles and defenses is up for grabs and promises to reshape strategic balances in ways that could profoundly alter the geostrategic positions of winners and losers alike. The stakes could not be higher. This new multipolar nuclear missile race poses dangers to deterrence stability and if left unchecked could over time provoke wider nuclear hedging or breakout. Current international controls are not up to the task of curtailing these dynamics and therefore need to be strengthened and augmented. Unfortunately, the prospects for this are remote in today's adversarial and shifting geopolitical and geostrategic international environment. Putting viable controls in place will therefore require Washington and its allies to play a long game that focuses on bending the incentives for China, Russia, and others to accept the need for new arms control constraints and nonproliferation guardrails. This will not be easy, and it will not happen quickly.

The nonproliferation and arms control ideas suggested here are by no means fully formed proposals. Most of these ideas have been bouncing around for years or even decades without attracting much interest. Modest and pragmatic options along these lines are not exciting to those seeking bold disarmament solutions. At the same time, even the most modest of them are still relatively far-fetched as China and others eschew negotiations. At most these are examples of seeds that could be planted and tended to see which ones might take root as geopolitical circumstances evolve over the coming decade.

Ultimately, solutions will not come from clever proposals. They will come instead from the realignment of hardnosed calculations across various capitals. The most important aspect of a new US approach to arms control and nonproliferation is therefore to achieve and utilize negotiating leverage. In practice this means demonstrating to China and Russia that they will gain nothing by arms racing against the United States and its allies and demonstrating to those allies that America has their back—all the while offering reasonable and pragmatic proposals to control the new nuclear arms race, temper hedging, and prevent new proliferation spirals.

CHAPTER 8:

ARMS CONTROL AMONG RIVALS

John D. Maurer

After decades of small wars and counterinsurgency, the United States is refocusing on great power rivalry with China and Russia.[555] This rivalry has profound implications for all aspects of defense planning, especially arms control negotiations.[556]

---

555. While the phrase "great power competition" may soon be replaced, geopolitical rivalry between the United States, China, and Russia will remain an enduring feature of international politics. On great power rivalry, see: "Summary of the 2018 National Defense Strategy of the United States of America: Sharpening the American Military's Competitive Edge," Department of Defense, 19 January 2018, https://dod.defense.gov/Portals/1/Documents/pubs/2018-National-Defense-Strategy-Summary.pdf, and Elbridge A. Colby and A. Wess Mitchell, "The Age of Great-Power Competition: How the Trump Administration Refashioned American Strategy," *Foreign Affairs*, January/February 2020, https://www.foreignaffairs.com/articles/2019-12-10/age-great-power-competition. On the language of great power rivalry, see: Zack Cooper, "Bad idea: 'Great power competition' terminology," Defense360, 1 December 2020, https://defense360.csis.org/bad-idea-great-power-competition-terminology/, and Robert C. O'Brien, "*A Free and Open Indo-Pacific*," White House, 5 January 2021, https://www.whitehouse.gov/wp-content/uploads/2021/01/OBrien-Expanded-Statement.pdf.

556. Maggie Tennis and Strobe Talbott, "Jettisoning arms control endangers America's edge in great-power politics," Brookings, 26 July 2019, https://www.brookings.edu/blog/order-from-chaos/2019/07/26/jettisoning-arms-control-endangers-americas-edge-in-great-power-politics/; John D. Maurer, "Post-INF great power arms control," *Real Clear Defense*, 17 September 2019, https://www.realcleardefense.com/articles/2019/09/17/post_inf_great_power_arms_control_114747.html; Christopher Ashley Ford, "The Politics of Arms Control: Getting Beyond Post-Cold War Pathologies and Finding Security in a Competitive Environment," Remarks, International Institute of Strategic Studies,

Many analysts worry that renewed great power rivalry will mean the end of arms control as a tool of national security policy.[557] These analysts are correct that changes in the international security environment will require significant deviation from the autopilot policies that guided American arms control policy for the last thirty years. Arms control in a world of great power rivalry will instead be *competitive arms control*. Great powers will use negotiations to promote their military advantages, the better to prevail over their rivals. If the United States does not wish to be left behind in this new world of competitive negotiation, then it must begin preparing to integrate arms limitation into its long-term strategy *now*.

In facing great power rivals, arms control can contribute to American security in three ways. First, the United States should avoid limiting weapons technologies where it enjoys decisive advantages, like missile defenses. Over the longer term, as adversaries diversify their own capabilities and catch up, limitation may become more desirable, but in the short term the United States gains little by trading away areas of strength. Second, the United States should focus on ensuring *equal* limitation of forces in which it can compete effectively with adversaries, but cannot predominate, like strategic nuclear forces. Over the longer term equal agreements will enhance the United States' major and durable geopolitical advantages. Third, the United States should accept adversary advantages in categories of weapons in which the United States does not wish to compete, like theater nuclear weapons. In these categories, the United States has little to lose to conceding to adversary demands, especially if they

---

London, 11 February 2020, https://www.state.gov/the-psychopolitics-of-arms-control/.

557. Eugene Rumer, "A Farewell to Arms… Control," Carnegie Endowment for International Peace, 17 April 2018, https://carnegieendowment.org/2018/04/17/farewell-to-arms-.-.-.-control-pub-76088; Greg Thielmann, "Are We Approaching the End of the Arms Control Era?" Remarks, Grinnell College, Grinnell, IA, 1 April 2019, https://www.armscontrol.org/blog/2019-04-01/we-approaching-end-arms-control-era; Linton F. Brooks, "The End of Arms Control?" *Daedalus* 149 (2), 2020, 84-100, https://doi.org/10.1162/DAED_a_01791; Ulrich Kühn, "Why Arms Control Is (Almost) Dead," Carnegie Endowment for International Peace Europe, 5 March 2020, https://carnegieeurope.eu/strategiceurope/81209.

are linked to adversary concessions in other more important areas.

To maximize advantages from the above three approaches over the longer term, American leaders will need to pursue three important tasks in the short term. The first is to compete vigorously with rivals in certain military-technical domains. Only a strong defense program will provide the leverage necessary to bring rivals to the negotiating table. The second is to integrate arms limitation negotiations into American strategy, along the lines described above. Finally, the United States must retain and expand its hard-won geopolitical advantages, especially its network of alliances and security partners throughout Eurasia. Arms control can support an effective long-term strategy, but it cannot substitute for it: even very successful negotiations will do little to promote American security if the fundamentals, such as alliance security, are allowed to crumble.

Past Competitive Arms Control

While arms control theory predicts that countries will use negotiations to advance their mutual interests in reducing the threat from inherently dangerous technologies, the historical record suggests that, in practice, arms control negotiations are often competitive.[558] Rival great powers move through a cycle of arms control negotiations: the contest for new weapons drives states to negotiate and, in turn, negotiations shape future rounds of rivalry. States manipulate this cycle to advance their own particular aims over time; they will deploy weapons aimed at shaping future negotiations in ways that benefit them, and they will negotiate agreements that shape future rounds to their advantage. For example, the great naval powers of the interwar period employed arms limitation negotiations to stall adversary naval construction, while pursuing transformative technologies like aircraft and submarines that each party hoped would

558. John D. Maurer, "The Purposes of Arms Control," *Texas National Security Review* 2, 1 (2018), 9-27, http://dx.doi.org/10.26153/tsw/870; Brendan Rittenhouse Green, *The Revolution that Failed: Nuclear Competition, Arms Control, and the Cold War* (New York: Cambridge University Press, 2020), 48-65.

prove decisive in future conflict.[559] Similarly, the United States used arms control negotiations with the Soviet Union during the Cold War to advance its "offset strategy," limiting the size of Soviet forces while allowing the United States to leap ahead in weapon accuracy and reliability.[560]

How do great powers use arms limitation negotiations for advantage? The most basic way is by deciding which weapons to limit, and which weapons to leave unrestrained. Great powers that enjoy a substantial advantage over their rivals in a certain type of weapon are unlikely to negotiate its limitation, and will instead prefer to exploit that advantage as much as possible. Similarly, great powers that are dramatically behind in an important weapons technology will generally prefer to rectify that major disadvantage, rather than ratify and extend their weakness into the future. Negotiations will instead focus primarily on technologies which both great powers are capable of producing and exploiting and whose ultimate benefit to either side is therefore ambiguous. Under those circumstances, great powers will seek to use negotiation to bolster their edge in that given technology.

Historically, we can see that great powers have engaged in negotiations selectively to promote their military-technical advantages. During the naval arms limitation negotiations of the 1920s and 1930s, for example, the United States and Great Britain preferred stricter limitations on asymmetric weapons technologies like submarines, which were potentially detrimental to their power-projection capabilities. Yet the Japanese, French, and Italians would not agree to limit this technology, which they believed could serve their interests in offsetting the battleship advantages of other countries.[561]

---

559. Emily O. Goldman, *Sunken Treaties: Naval Arms Control Between the Wars* (University Park, PA: The Pennsylvania State University Press, 1994), 153-188.
560. John D. Maurer, "The Forgotten Side of Arms Control: Enhancing U.S. Competitive Advantage, Offsetting Enemy Strengths," *War on the Rocks*, 27 June 2018, https://warontherocks.com/2018/06/the-forgotten-side-of-arms-control-enhancing-u-s-competitive-advantage-offsetting-enemy-strengths/.
561. Joel Ira Holwitt, *"Execute Against Japan": The U.S. Decision to Conduct*

On the other hand, during the early Cold War the United States tried to freeze the strategic nuclear balance in a position of significant American numerical superiority. Unsurprisingly, Soviet leaders would not accept inferiority in such a crucial category of weapons, and continued their own buildup until they reached numerical parity with the United States in the late 1960s.[562]

Rivalrous great powers generally avoid arms control negotiations when they believe that unilateral measures will allow them to prevail over rivals, but when the future balance of capability is less certain they will use negotiations to promote their own strengths. The search for advantage through arms control often results in early rounds of negotiation in which each power demands obvious, unilateral advantages in number and quality of forces over its rival. Yet paradoxically negotiation can only promote one's relative advantage if the adversary consents to and honors the agreement. To win the adversary's assent while preserving one's own advantage, great powers employ two major strategies in their competitive arms control negotiations: *equality,* or *equivalence*. Under a strategy of "equality," a great power offers the adversary a seemingly fair, numerically equal agreement, with the tacit understanding that numerical equality of forces will enhance one's own fewer tangible advantages. Conversely, under a strategy of "equivalence," a great power demands agreements numerically weighted in its favor, which are justified as necessary to offset some larger advantage of its rival.

Whether a great power demands equality or equivalence in arms control negotiations depends on its specific combination of strengths and weaknesses beyond the military realm – economic, technological, and geopolitical. First, a state that is much weaker economically than its rival is more likely to offer negotiations premised on numerically equal forces, since it would be outpaced in an unlimited arms contest. On the other hand, an economically powerful state

---

Unrestricted Submarine Warfare (College Station, TX: Texas A&M University Press, 2009), 29-42.

562. James Cameron, *The Double Game: The Demise of America's First Missile Defense System and the Rise of Strategic Arms Limitation* (New York: Oxford University Press, 2018), 49-106.

is likely to demand a higher "equivalent" level of forces reflecting its economic strength. Second, a state that has more sophisticated weapons technology than its adversary is also likely to offer negotiations based on numerical equality, since this will force future contests into the qualitative domain. Conversely, a state that recognizes its technological inferiority will likely demand numerically larger equivalent forces to offset its qualitative disadvantages. Third, a state that enjoys major geopolitical advantages – physical security, easy access to a strategic theater, or many helpful local partners – will also generally offer to negotiate on the basis of equality, trusting that its geopolitical strengths will allow it to prevail. Countries that are less geopolitically secure or surrounded by many hostile states will conversely demand larger equivalent forces that reflect their malign environment.

Because the economic, technological, and geopolitical factors driving great power arms control negotiations reflect *relative* advantages, it is common for different great powers to prefer different approaches, with one calling for equality while the other demands equivalence. Much of the drama of previous negotiations has been bound up in the struggle to determine which approach will prevail, with the outcome often determined by how large the relative advantages of each power are. Take, for example, the naval arms limitation negotiations of the 1920s and 1930s. During this period, Japan's relative economic weakness and geopolitically secure position in the Western Pacific led it to demand *equality* in number of major warships with the other major naval powers, Great Britain and the United States. Japan's far weaker economy meant that it would be swamped in a naval arms race, something that Japanese leaders hoped to avoid. At the same time, Japan's isolated position in the Western Pacific made it difficult for adversaries to attack, while also giving it prime position to expand its colonial holdings in East Asia. An agreement establishing numerical equality of naval forces thus suited Japan's economic and geopolitical situation.

Yet given their economic strengths and geopolitical disadvantages *vis a vis* Japan, Britain and the United States would not settle for equal naval forces. Because Japan was a "one ocean" power while Britain and the United States were each "two ocean" powers, British and American leaders argued that they needed *more* forces than Japan. Put another way, British and American leaders argued that their geopolitical situations were worse than that of Japan, and so required unequal forces to ensure equivalent security – an *equivalent* approach. Japanese leaders were unhappy with this arrangement, but faced with the threat of a naval arms race against the much larger economies of *both* Britain and the United States, they reluctantly agreed to an unequal but "equivalent" level of forces, even though this meant that the Japanese Navy might end up outgunned in a future fight if the British or Americans concentrated all of their forces on the Pacific.[563]

During the Cold War, however, the United States abandoned its demands for arms control equivalence and instead insisted that negotiations be conducted on the principle of *equality* between the superpowers.[564] American leaders calculated that their network of alliances and advanced industrial base would allow them to produce and deploy qualitatively superior weapons, and that Soviet-American numerical equality would therefore advance American strength.[565] The Soviets insisted that arms control agreements reflect the Soviet Union's much more difficult security situation, surrounded as they were by American-backed allies and hostile former clients like China.[566] Soviet leaders agreed that American alliances

563. Goldman, *Sunken Treaties*, 111-152; John H. Maurer, "Arms Control and the Washington Conference," in *The Washington Conference, 1921-22: Naval Rivalry, East Asian Stability and the Road to Pearl Harbor*, eds. Erik Goldstein and John Maurer (New York: Frank Cass, 1994), 267-293.

564. "National Security Decision Memorandum 74," 31 July 1970, *Foreign Relations of the United States, 1969-1976, Volume XXXII: SALT I, 1969-1972*, ed. Erin R. Mahan (Washington, DC: United States Government Printing Office, 2010), Document 100, https://history.state.gov/historicaldocuments/frus1969-76v32/d100.

565. "Conversation Among President Nixon, the Joint Chiefs of Staff, Secretary of Defense Laird, and Others," 10 August 1971, *FRUS 1969-76,* Vol. XXXII, Document 190, https://history.state.gov/historicaldocuments/frus1969-76v32/d190.

566. Memo, Sonnenfedlt to Kissinger, 20 May 1970, *FRUS 1969-76*, Vol. XXXII, Document 77, https://history.state.gov/historicaldocuments/frus1969-76v32/d77.

and technology gave the United States a sizeable advantage, and so preferred that arms control grant them superiority in the number and size of weapons, which they argued would be equivalent.[567]

Yet Soviet demands for larger forces to offset their technological and geopolitical weaknesses ran into trouble given the much smaller Soviet economy. Initially, Soviet leaders tried to split the difference: they conceded equality in strategic weapons in the mid-1970s, but still demanded numerical advantage in theater-range weapons targeting American allies.[568] Only as their economy was truly crumbling, and facing a new generation of American missiles, did Soviet leaders finally agree to American demands for equal strategic and theater forces in the 1987 INF and 1991 START I Treaties.[569] Under the INF Treaty, the Soviets agreed to dismantle their land-based intermediate-range weapons while permitting the United States to retain its growing arsenal of similar sea-based weapons.[570] Similarly, under the START I Treaty the Soviets agreed to smaller, numerically equal strategic arsenals, even as the United States deployed a new generation of qualitatively-superior and highly-accurate Peacekeeper and

---

567. Aleksandr' G. Savel'yev and Nikolay N. Detinov, *The Big Five: Arms Control Decision-Making in the Soviet Union*, ed. Gregory Varhall, trans. Dimitriy Trenin (Westport, CT: Praeger, 1995), 10-11; John Hines, Ellis Mishulovich, and John Shull, *Soviet Intentions 1965-1985, Volume I: An Analytical Comparison of U.S.-Soviet Assessments During the Cold War* (McLean, VA: BDM Federal, 1995), 4-8, http://nsarchive.gwu.edu/nukevault/ebb285/.
568. Steven Pifer *et al*, *The Treaty on Intermediate-Range Nuclear Forces: History and Lessons* (Washington, DC: Brookings Institution, 2012), https://www.brookings.edu/research/the-treaty-on-intermediate-range-nuclear-forces-history-and-lessons-learned/.
569. "Treaty Between the United States of America and the Soviet Union of Socialist Republics on the Elimination of Their Intermediate-Range and Shorter Missiles (INF Treaty," United States Department of State, accessed 16 September 2020, https://2009-2017.state.gov/t/avc/trty/102360.htm; "START I at a Glance," *Arms Control Association*, February 2019, https://www.armscontrol.org/factsheets/start1.
570. John D. Maurer, "The Dual-Track Approach: A Long-Term Strategy for a Post-INF Treaty World," *War on the Rocks*, 10 April 2019, https://warontherocks.com/2019/04/the-dual-track-approach-a-long-term-strategy-for-a-post-inf-treaty-world/.

Trident II missiles.[571]

Future Competitive Arms Control

As the old Russian-American bargains break down and new rivals emerge, the big question is how previous dynamics will map onto future arms control negotiations. First, we should expect little from negotiations where one side enjoys an obvious advantage. Second, we should expect countries that *do* engage in negotiation to wrestle with the question of equality and equivalence. As before, we should expect great powers to bargain hard at the negotiating table for whatever advantages they can wring from the adversary. Yet countries that are weaker economically or stronger technologically and geopolitically should be more willing to offer numerically "equal" negotiations that compensate for their economic weakness or allow their technological and geopolitical advantages freest reign. Conversely, countries that are economically stronger or technologically and geopolitically weaker will continue to demand numerically greater "equivalent" forces, reflecting their economic advantages and/or offsetting their qualitative weaknesses.

What does this all mean for the United States? Given its massive advantages, the United States should engage in arms limitation negotiations selectively, and avoid limiting areas of significant strength. When it *does* negotiate, the United States should still generally prefer numerically *equal* arms limitation. For decades the United States' economic and technological advantages have dwarfed those of rivals, allowing it to prevail qualitatively even in contests where the number of weapons was limited. As such, the United States should always be on the lookout for opportunities to head off arms racing and further extend these American advantages.

Yet even as rivals like China close the economic and technologi-

571. Maurer, "The Forgotten Side of Arms Control," https://warontherocks.com/2018/06/the-forgotten-side-of-arms-control-enhancing-u-s-competitive-advantage-offsetting-enemy-strengths/.

cal gap and some traditional American advantages wane, the United States can build on the advantages of its unique geopolitical position, including its relative geographic isolation from its rivals and its large network of alliances throughout Eurasia. America's strategic bastion in the Western Hemisphere means that its Eurasian rivals will always pose a greater threat to each other than the United States does, hindering their ability to coordinate effectively beyond diplomatic posturing.[572] At the same time, the United States' vast network of allies complicates adversaries' defense planning and provides the United States with strategic options that adversaries lack.[573] Where it cannot retain significant superiority in numbers, the United States should stand firm on numerical equality in arms control negotiations, confident that its durable advantages will allow it to prevail over its rivals. For this very reason, we should also expect Chinese and Russian leaders to resist "equal" arms control proposals, and instead insist on numerical advantage in weapons they perceive as offsetting American geopolitical strengths.

As the United States considers its future arms control options, it will also have to contend with the shift from the bipolar world of

---

572. Paul Stronski and Nicole Ng, "Cooperation and Competition: Russia and China in Central Asia, the Russian Far East, and the Arctic," *Carnegie Endowment for International Peace,* 28 February 2018, https://carnegieendowment.org/2018/02/28/cooperation-and-competition-russia-and-china-in-central-asia-russian-far-east-and-arctic-pub-75673; Tanvi Madan, "Between a Cold War Ally and an Indo-Pacific Partner: India's U.S.-Russia Balancing Act," *War on the Rocks,* 16 October 2018, https://warontherocks.com/2018/10/between-a-cold-war-ally-and-an-indo-pacific-partner-indias-u-s-russia-balancing-act/; Pavel K. Baev, "The limits of authoritarian compatibility: Xi's China and Putin's Russia," The Brookings Institution, June 2020, https://www.brookings.edu/research/the-limits-of-authoritarian-compatibility-xis-china-and-putins-russia/.

573. Mira Rapp-Hooper, "Saving America's Alliances: The United States Still Needs the System That Put It On Top," *Foreign Affairs*, March/April 2020, https://www.foreignaffairs.com/articles/united-states/2020-02-10/saving-americas-alliances; David A. Wemer, "Defense Secretary unveils a new strategy for bolstering allies and partnerships in an era of great-power competition," *New Atlanticist*, 20 October 2020, https://www.atlanticcouncil.org/blogs/new-atlanticist/defense-secretary-unveils-a-new-strategy-for-bolstering-allies-and-partnerships-in-an-era-of-great-power-competition/.

the Cold War to the multipolar world of the 21st century. Yet as the naval arms limitation negotiations of the interwar period demonstrate, the basic principles of competitive arms control can and do apply in a multipolar contest – countries will still negotiate only when it suites them, and then structure those negotiations towards equality or equivalence depending on their relative strengths. Of course, multilateral arms limitation negotiations will likely remain a long-term aspiration. The conventional wisdom is that multilateral negotiations are harder to conduct and conclude than bilateral ones, given the added difficulties of satisfying everyone's demands and the added points of failure in additional ratification processes.[574]

As it approaches multilateral arms control negotiations, the most important question for the United States is what role American allies and security partners will play. On the one hand, the United States could exclude allies from negotiations, and limit talks to only the United States and its rivals. This was the approach that the United States pursued during the Cold War, when most of the key arms limitation negotiations took place between the superpowers. This approach can yield tremendous benefits, since by demanding equality with its rivals the United States can effectively augment the strength of its allied partners, whose forces are not counted. On the other hand, this approach demands that the United States shoulder many of the burdens of rivalry and negotiation by itself, a potentially costly proposition. It also poses challenges for alliance cohesion, as allies and partners will have to trust that the United States is representing their best interests at the negotiating table. Canny adversaries may try to exploit this potential mistrust, much as the Soviets did during the Cold War.[575]

Alternatively, the United States could try to "seed" the negotiat-

---

574. Larry Crump and I. William Zartman, "Multilateral Negotiation and the Management of Complexity," *International Negotiation* 8, 1 (2003), 1-5, https://doi.org/10.1163/138234003769590640.

575. Memo, Lynn and Sonnenfeldt to Kissinger, 1 July 1970, *Foreign Relations of the United States, 1969-1976, Volume XXXII: SALT I, 1969-1972*, ed. Erin Mahan (Washington, DC: United States Government Printing Office, 2010), Document 86, https://history.state.gov/historicaldocuments/frus1969-76v32/d86.

ing table with its own security partners as well as rivals. Including friendly states in negotiations could allow the American "side" to generate greater leverage in talks. For example, the United States was able to impose its preferred 5:3 ratio on Japan during interwar naval talks in large part because the British came to support its stance, threatening Japan with an overwhelming naval coalition arrayed against it. On the other hand, including security partners in talks could allow rivals to demand more "equivalent" forces, such as a Russian demand to equal all of NATO's strategic nuclear capabilities, not just those of the United States. Coordinating allied negotiating strategy would also pose political and diplomatic challenges to ensure that rivals could not divide the United States and its partners against each other.

In the shorter term, however, continued bilateral arrangements between the great powers will still have an important role in multilateral rivalry. First, agreements that enable two powers to cooperate more effectively against common rivals will still have relevance, like the United States and India's growing détente on nuclear issues.[576] Second, agreements that enable two powers to compete with each other at lower costs will also remain relevant. For example, continued Russian-American bilateral strategic dialogue could be important in redirecting resources from both countries towards competing with China.[577] Third, agreements might shape *where* rivals clash geographically. For example, the United States and Russia could conclude agreements covering only Europe but not Asia.[578] Finally, agreements might have implications from *when* rivalry takes place,

576. Mohan Malik, "China Responds to the U.S.-India Nuclear Deal," *China Brief* 6, 7 (2006), https://jamestown.org/program/china-responds-to-the-u-s-india-nuclear-deal/; Amitai Etzioni, "The Darker Side of the U.S.-India Nuclear Deal," *The Diplomat*, 13 February 2015, https://thediplomat.com/2015/02/the-darker-side-of-the-u-s-india-nuclear-deal/.
577. John D. Maurer, "America's strategic interest in New START," *AEIdeas*, 5 November 2019, https://www.aei.org/foreign-and-defense-policy/americas-strategic-interest-in-new-start/.
578. "The Post-INF Treaty Crisis: Background and Next Steps," *Arms Control Association*, 7 August 2019, https://www.armscontrol.org/issue-briefs/2019-08/post-inf-treaty-crisis-background-next-steps.

allowing countries to accelerate or delay key phases of arms racing. For example, the United States might seek to delay racing with the Russians while their technological base declines or with the Chinese while their demographic and environmental challenges multiply.[579]

Scenarios for Future Great Power Arms Control

The return of great power rivalry means that arms control negotiations are unlikely to succeed in the short term, especially as the legacy Cold War agreements are outstripped by new circumstances and while the United States belatedly retools for genuine strategic rivalry. Yet, over the longer term, new arms limitation negotiations between the great powers are very likely. Under what circumstances might we see renewed negotiations over limiting great power armaments? What might these negotiations look like? In the following section, I consider possible limitation of five categories of weapons, describing the current and likely future balance of capabilities among the great powers, and extrapolating what sorts of arms control policies the United States should adopt.

Strategic Nuclear Weapons: Strategic nuclear weapons are an area where the United States is strong, but not dominant. The United States and Russia enjoy roughly similar strategic nuclear arsenals, but the relevant Chinese and Indian forces are an order of magnitude smaller.[580] Chinese and Indian nuclear forces are likely to expand over the long term, though observers remain uncertain about how large their arsenals will ultimately become.[581] Chinese and Indian

---

579. Apurva Sanghi and Shahid Yusuf, "Russia's Uphill Struggle with Innovation," The World Bank, 17 September 2018, https://www.worldbank.org/en/news/opinion/2018/09/17/russias-uphill-struggle-with-innovation; Michael Beckley, "China's Economy Is Not Overtaking America's," *Journal of Applied Corporate Finance* 32, 2 (2020), 10-23, https://doi.org/10.1111/jacf.12401.
580. "Nuclear Weapons: Who Has What at a Glance," *Arms Control Association*, August 2020, https://www.armscontrol.org/factsheets/Nuclearweaponswhohaswhat.
581. For arguments that China and India will build moderately in coming years see: Hans M. Kristensen and Matt Korda, "Chinese nuclear forces, 2019," *Bul-*

leaders will need to weigh the benefits of larger strategic nuclear arsenals against the costs of expanding those arsenals, especially as they also race to build up theater-range missile capabilities.

At the same time, China, Russia, and the United States are all actively working to increase the quality of their strategic nuclear forces, and continued advances in sensing capabilities and weapon accuracy will increase counterforce capabilities. While weapon mobility and concealment will continue to complicate counterforce targeting, over the longer term improvements in sensing techniques and data processing will continue to erode the survivability of second-strike forces in the 2030s and beyond.[582] While retaining accurate missile capabilities will remain important, the locus of strategic nuclear rivalry is likely to shift into the information domain, as great powers increasingly compete to amass the targeting data necessary to attack adversary forces.[583] The search for targeting information is likely to take new and more active forms, including intrusive electronic and cyber measures, with potentially destabilizing consequences.[584]

---

letin of the Atomic Scientists 75, 4 (2019), 171-178, https://doi.org/10.1080/00963402.2019.1628511, and Hans M. Kristensen and Matt Korda, "Indian nuclear forces, 2020," *Bulletin of the Atomic Scientists* 76, 4 (2020), 217-225, https://doi.org/10.1080/00963402.2020.1778378. For arguments that China especially may expand its arsenal significantly, see: Michael Mazza and Henry Sokolski, "China's Nuclear Arms Are a Riddle Wrapped in a Mystery," *Foreign Policy*, 13 March 2020, https://foreignpolicy.com/2020/03/13/china-nuclear-arms-race-mystery/; and Peter Huessy and Bradley A. Thayer, "China's nuclear developments reflect its growing ambition," *The Hill*, 7 November 2019, https://thehill.com/opinion/national-security/468972-chinas-nuclear-developments-reflect-its-growing-ambition.

582. Austin Long and Brendan Green, "Stalking the Secure Second Strike: Intelligence, Counterforce, and Nuclear Strategy," *Journal of Strategic Studies* 38, 1-2 (2015), 38-73, https://doi.org/10.1080/01402390.2014.958150; Keir A. Lieber and Daryl G. Press, "The New Era of Counterforce: Technological Change and the Future of Nuclear Deterrence," *International Security* 41, 4 (2017), 9-49, https://doi.org/10.1162/ISEC_a_00273.

583. Paul Bracken, "The Intersection of Cyber and Nuclear War," *The Strategy Bridge*, 17 January 2017, https://thestrategybridge.org/the-bridge/2017/1/17/the-intersection-of-cyber-and-nuclear-war.

584. Bryan Clark, "2019 Forecast: Hard Choices on Invisible Warfare," *Breaking Defense*, 4 January 2019, https://breakingdefense.com/2019/01/2019-fore-

Given the uncertainties surrounding this information contest, the United States and its great power rivals are likely to hedge against major counterforce breakthroughs by retaining sizeable nuclear arsenals to bolster their survivability.[585]

The need to maintain sufficiently large strategic nuclear forces to improve their survival will have significant implications for future arms limitation efforts. For the United States and Russia, the search for survivability in numbers will likely mean an end to the decades-long process of mutual, agreed reduction of their nuclear arsenals.[586] Further marginal reduction of American or Russian forces may still occur, depending on the fiscal situation of the two countries, but such reductions are unlikely to occur according to a mutual negotiated plan, given the potential for serious future counterforce threats.[587]

---

cast-hard-choices-on-invisible-warfare/; Niall Firth, "How to fight a war in space (and get away with it)," *MIT Technology Review*, 26 June 2019, https://www.technologyreview.com/2019/06/26/725/satellite-space-wars/; James Johnson and Eleanor Krabil, "AI, Cyberspace, and Nuclear Weapons," *War on the Rocks*, 31 January 2020, https://warontherocks.com/2020/01/ai-cyberspace-and-nuclear-weapons/.

585. Long and Green, "Stalking the Secure Second Strike;" Lieber and Press, "The New Era of Counterforce."

586. Charles L. Glaser, "The Instability of Small Numbers Revisited: Prospects for Disarmament and Nonproliferation," in *Rebuilding the NPT Consensus*, ed. Michael May (Stanford, CA: Center for International Security and Cooperation, 2008), 217-230, https://cisac.fsi.stanford.edu/publications/rebuilding_the_npt_consensus; Thomas C. Schelling, "A world without nuclear weapons?" *Daedalus* 138, 4 (2009), 124-129, https://doi.org/10.1162/daed.2009.138.4.124.

587. For recent discussions of reducing spending on American nuclear forces, see: William D. Hartung, "Now isn't the time to push for nuclear modernization," *Defense News*, 21 April 2020, https://www.defensenews.com/opinion/commentary/2020/04/21/now-isnt-the-time-to-push-for-nuclear-modernization/; Steven Pifer, "How COVID-19 might affect US nuclear weapons and planning," Brookings, 18 May 2020, https://www.brookings.edu/blog/order-from-chaos/2020/05/18/how-covid-19-might-affect-us-nuclear-weapons-and-planning/; and Kingston Reif, "Debating US nuclear spending in the age of the coronavirus," *Bulletin of the Atomic Scientists*, 10 June 2020, https://thebulletin.org/2020/06/debating-us-nuclear-spending-in-the-age-of-the-coronavirus/. On Russia's fiscal-military challenges, see: Susanne Oxenstierna, "Russia's defense spending and the economic decline," *Journal of Eurasian Studies* 7, 1

For China and India, bolstering force survivability will likely mean expanding and diversifying their nuclear forces into full triads.[588]

American strategic nuclear arms limitation policy will have to evolve significantly in the coming years as the Chinese arsenal grows in size. For decades, the smaller size of the Chinse arsenal has allowed the United States to retain roughly equal numbers of strategic launchers with *both* of its major rivals, China and Russia. In recent years, the United States has maintained 655 strategic missiles and bombers, while Russia and China have deployed a combined total of 621.[589] The United States should do everything it can to sustain this enviable position as long as possible by limiting the size of Chinese strategic nuclear forces, even if it means conceding

---

(2016), 60-70, https://doi.org/10.1016/j.euras.2015.06.001; Pavel Baev, "Russian Nuclear Modernization and Putin's Wonder-Missiles: Real Issues and False Posturing," Ifri, August 2019, https://www.ifri.org/sites/default/files/atoms/files/baev_russian_nuclear_modernization_2019.pdf; Keith Crane, Olga Oliker, and Brian Nichiporuk, *Trends in Russia's Armed Forces: An Overview of Budgets and Capabilities* (Santa Monica, CA: The RAND Corporation, 2019), https://www.rand.org/content/dam/rand/pubs/research_reports/RR2500/RR2573/RAND_RR2573.pdf.

588. On Chinese nuclear modernization, see: Lt. Gen. Robert P. Ashley, Jr., "Russian and Chinese Nuclear Modernization Trends," Remarks, Hudson Institute, 29 May 2019, https://www.dia.mil/News/Speeches-and-Testimonies/Article-View/Article/1859890/russian-and-chinese-nuclear-modernization-trends/; Kristensen and Korda, "Chinese nuclear forces, 2019," https://doi.org/10.1080/00963402.2019.1628511; *Military and Security Developments Involving the People's Republic of China, 2020: Annual Report to Congress* (Washington, DC: Office of the Secretary of Defense, 2020), https://media.defense.gov/2020/Sep/01/2002488689/-1/-1/1/2020-DOD-CHINA-MILITARY-POWER-REPORT-FINAL.PDF. On Indian nuclear modernization, see: Kristensen and Korda, "Indian nuclear forces, 2020," https://doi.org/10.1080/00963402.2020.1778378.

589. The American and Russian arsenals are monitored under the New START Treaty ("New START Treaty Aggregate Numbers of Strategic Offensive Arms," Department of State, 1 March 2020, https://www.state.gov/new-start-treaty-aggregate-numbers-of-strategic-offensive-arms-13/). As of early 2019, China was estimated to deploy around 90 intercontinental ballistic missiles and 46 submarine-launched ballistic missiles (Kristensen and Korda, "Chinese nuclear forces, 2019," https://doi.org/10.1080/00963402.2019.1628511).

superiority to China and Russia in areas where the United States is weaker, like theater nuclear weapons. One important tool for limiting the scope of the Chinese nuclear arsenal could be greater controls on the production of nuclear-weapons-grade fissile material, of which the United States maintains large stockpiles and which China is working to expand.[590] American negotiators might gain greater leverage in limiting fissile material if American allies like Japan and South Korea also agreed to such limitations, which would clearly be in Beijing's interest.[591] Barring some limitation of Chinese nuclear production, however, the United States is unlikely to sustain its "two-power standard" in strategic nuclear weapons in an unlimited contest over the longer term. As the Chinese arsenal grows, the United States may have to fall back to a position of "equality" with *each* of its rivals' arsenals – that is, allowing Russia and China to each have as many strategic nuclear weapons as the United States.

While the United States should do what it can to forestall the growth of the Chinese nuclear arsenal, the number of Chinese and Russian weapons is likely to be less decisive in future great power rivalry than the quality of information guiding them. In this regard, the

---

590. On Chinese fissile material production, see: Henry Sokolski, "China's Nuclear Weapons and Fissile Materials Holdings: Uncertainties and Concerns," Testimony before the U.S.-China Economic and Security Review Commission. 26 March 2012, http://www.npolicy.org/article.php?aid=1168&tid=4, and Bill Gertz, "Exclusive: China's 'secretive, crash' nuclear buildup revealed," *Washington Times*, 12 November 2020, https://www.washingtontimes.com/news/2020/nov/12/china-expanding-nuclear-arms-plants-revealed/. On limiting fissile material production, see: James M. Acton, Thomas Macdonald, and Pranay Vaddi, "Revamping Nuclear Arms Control: Five Near-Term Proposals," Carnegie Endowment for International Peace, 14 December 2020, https://carnegieendowment.org/2020/12/14/revamping-nuclear-arms-control-five-near-term-proposals-pub-83429, and Tong Zhao, "Practical Ways to Promote U.S.-China Arms Control Cooperation," Carnegie-Tsinghua Center for Global Policy, 7 October 2020, https://carnegietsinghua.org/2020/10/07/practical-ways-to-promote-u.s.-china-arms-control-cooperation-pub-82818.
591. Michael Mazza and Henry Sokolski, "America Must Dissuade China's Interest in Nuclear Arms," *The National Interest*, 14 December 2019, https://nationalinterest.org/feature/america-must-dissuade-chinas-interest-nuclear-arms-105117.

United States is likely to enjoy significant and durable advantages in access to large volumes of high-quality information. While China and Russia each have significant information gathering capabilities and strong incentives to cooperate against the United States, their efforts to pool and share scientific and informational resources are hampered by their own track records of espionage and intellectual property theft, including from each other.[592] By comparison, the United States not only enjoys its own powerful intelligence capabilities, but also the deepest intelligence cooperation with allies and partners around the world, providing an unparalleled reservoir of information for future targeting.[593] Even in a more numerically equal world, the United States would thus still enjoy greater qualitative informational strategic nuclear capability than its major rivals.[594]

Although the United States has a much deeper bench of intelligence partners than China or Russia, it might still want to consider negotiations to limit active cyber intrusions against nuclear forces. Aside from the possibility that such intrusions could be highly destabilizing, they are also an area where China and Russia seem to have significant capabilities, as gauged by the high-profile hacking campaigns of recent years.[595] While the United States has significant cy-

---

592. Samuel Bendett and Elsa Kania, "Resilience of Sino-Russian High-Tech Cooperation," *War on the Rocks*, 12 August 2020, https://warontherocks.com/2020/08/the-resilience-of-sino-russian-high-tech-cooperation/.
593. Noah Barkin, "Exclusive: Five Eyes intelligence alliance builds coalition to counter China," *Reuters*, 12 October 2018, https://www.reuters.com/article/us-china-fiveeyes/exclusive-five-eyes-intelligence-alliance-builds-coalition-to-counter-china-idUSKCN1MM0GH; Vivek Raghuvanshi, "India, US sign intel-sharing agreement amid tension with neighboring China," *Defense News*, 28 October 2020, https://www.defensenews.com/space/2020/10/28/india-us-sign-intel-sharing-agreement-amid-tension-with-neighboring-china/; "RESOLVED: Japan is Ready to Become a Formal Member of the Five Eyes," *CSIS Debating Japan* 3, 8 (December 2020), https://www.csis.org/analysis/resolved-japan-ready-become-formal-member-five-eyes.
594. Paul Bracken, "In the Second Nuclear Age, Information Advantage Defines the Balance of Power," *Yale Insights*, 11 August 2020, https://insights.som.yale.edu/insights/in-the-second-nuclear-age-information-advantage-defines-the-balance-of-power.
595. Jordan Robertson and Michael Riley, "The Big Hack: How China Used a

ber capabilities of its own, the relative balance of these capabilities is difficult for the public to assess.[596] If American leaders calculate that Chinese and Russian cyber intrusions help them compensate for their lack of significant multilateral intelligence sharing, however, then negotiations to limit active cyber measures, especially against nuclear forces, would ultimately play to American strengths. Of course, cyber arms control faces significant obstacles, including the challenges of specifying the capabilities or activities to be limited *and* the difficulties of verifying restraints on covert cyber activities.[597] Limiting the cyber-attacks against nuclear forces would thus likely involve less formal commitments by the major powers to show mutual restraint in cyber intrusions, with verification ensured primarily by unilateral technical means.[598] If such informal agreements could bolster American informational advantages, then they would be well worth pursuing.

Long-Range Conventional Weapons: When it comes to theater conventional weapons (up to 5,500 km in range), the United States

---

Tiny Chip to Infiltrate U.S. Companies," *Bloomberg*, 4 October 2018, https://www.bloomberg.com/news/features/2018-10-04/the-big-hack-how-china-used-a-tiny-chip-to-infiltrate-america-s-top-companies; David E. Sanger, "Russian Hackers Broke Into Federal Agencies, U.S. Officials Suspect," 13 December 2020, https://www.nytimes.com/2020/12/13/us/politics/russian-hackers-us-government-treasury-commerce.html.

596. Joshua Rovner, "More Aggressive and Less Ambitious: Cyber command's Evolving Approach," *War on the Rocks,* 14 September 2020, https://warontherocks.com/2020/09/more-aggressive-and-less-ambitious-cyber-commands-evolving-approach/.

597. Erica D. Borghard and Shawn W. Lonergan, "Why Are There No Cyber Arms Control Agreements?" *Council on Foreign Relations*, 16 January 2018, https://www.cfr.org/blog/why-are-there-no-cyber-arms-control-agreements; Andrew Futter, "What does cyber arms control look like? Four principles for managing cyber risk," European Leadership Network, June 2020, https://www.europeanleadershipnetwork.org/wp-content/uploads/2020/06/Cyber-arms-control.pdf.

598. Christopher Ashley Ford, "Rules, Norms, and Community: Arms Control Discourses in a Changing World," Remarks, European Union Conference on Nonproliferation, Brussels, 13 December 2019, https://www.state.gov/rules-norms-and-community-arms-control-discourses-in-a-changing-world/.

is in an ambiguous position. Today, the United States is by far the world leader in sea- and air-based theater conventional weapons, with an arsenal of thousands of sea- and air-launched cruise missiles, while China has the world's greatest ground-based capabilities, including nearly a thousand truck-launched medium- and intermediate-range ballistic missiles.[599] This current balance is unlikely to be durable, however, as the end of the INF Treaty frees the United States to pursue its own ground-based capabilities, while the rapid buildup of Chinese and Indian naval forces challenge America's long-standing naval preeminence.[600] Yet assuming that it can keep its allies onboard, the United States will enjoy durable advantages in theater conventional strikes given its large network of alliances throughout Europe and Asia. American weapons deployed on and near allied territory can strike adversaries like China or Russia directly, while Chinese and Russian intermediate-range weapons can strike American allies and bases, but not the United States itself.[601] Furthermore, that same network of alliances constrains the ability of Chinese or Russian naval forces to project power outside of their home regions, further worsening the theater conventional balance.[602] Over the longer-term, the contest for theater conventional weapons is likely to cause greater harm to the security of Eurasian powers

599. Eric Heginbotham *et al*, *The U.S.-China Military Scorecard: Forces, Geography, and the Evolving Balance of Power, 1996-2017* (Santa Monica, CA: The RAND Corporation, 2015), https://www.rand.org/pubs/research_reports/RR392.html.
600. Sydney J. Freedberg, Jr., "What Weapons Will the US Build After The INF Treaty," *Breaking Defense*, 22 October 2018, https://breakingdefense.com/2018/10/what-weapons-will-the-us-build-after-the-inf/; Andrey Baklitskiy, "What the End of the INF Treaty Means for China," Carnegie Moscow Center, 12 February 2019, https://carnegie.ru/commentary/80462; Franz-Stefan Gady, "India Test Fires Intermediate-Range, Submarine-Launched Ballistic Missile," *The Diplomat*, 20 January 2020, https://thediplomat.com/2020/01/india-test-fires-intermediate-range-submarine-launched-ballistic-missile/.
601. Maurer, "The Dual-Track Approach," https://warontherocks.com/2019/04/the-dual-track-approach-a-long-term-strategy-for-a-post-inf-treaty-world/.
602. Michael Beckley, "The Emerging Military Balance in East Asia: How China's Neighbors Can Check Chinese Naval Expansion," *International Security* 42, 2 (2017), 78-119, https://www.mitpressjournals.org/doi/pdf/10.1162/ISEC_a_00294.

than to the United States which, even in the missile age, enjoys an advantageous isolation in the Western Hemisphere.

While the technical characteristics of theater-range weapons favor the United States and its network of allies, reliance on security partners poses significant political challenges that rivals like China and Russia do not face. American theater missile deployments require partners to agree to host American missiles, a difficult political proposition. Indeed, in the immediate aftermath of the INF Treaty, allied governments rejected or downplayed the possibility of hosting new American missiles, leading many analysts to conclude that American allies would never agree to host such weapons.[603] Yet as the INF Treaty recedes into the past and military rivalry continues unabated, many American partners are taking a second look at the potential utility of theater-range conventional missiles, providing opportunities for fruitful collaboration.[604] Ironically, Chinese and Russian efforts to disrupt cooperation between the United States and its al-

603. On allied responses, see: "Australia won't host U.S. missiles, prime minster says," *Reuters*, 5 August 2019, https://www.reuters.com/article/us-australia-usa-missiles/australia-wont-host-u-s-missiles-prime-minister-says-idUSKCN1U-V0IB; Robin Harding and Lionel Barber, "Japan sounds warning on China's growing military might," *Financial Times*, 31 October 2019, https://www.ft.com/content/eb5191ee-fbbf-11e9-a354-36acbbb0d9b6; Kingston Reif, "AS INF Treaty Falls, New START Teeters," *Arms Control Association*, March 2019, https://www.armscontrol.org/act/2019-03/news/inf-treaty-falls-new-start-teeters. For initial assessments, see: Pranay Vaddi, "Leaving the INF Treaty Won't Help Trump Counter China," Carnegie Endowment for International Peace, 31 January 2019, https://carnegieendowment.org/2019/01/31/leaving-inf-treaty-won-t-help-trump-counter-china-pub-78262; Tom Countryman and Kingston Reif, "Intermediate-Range Missiles are the Wrong Weapon for Today's Security Challenges," *War on the Rocks*, 13 August 2019, https://warontherocks.com/2019/08/intermediate-range-missiles-are-the-wrong-weapon-for-todays-security-challenges/; "The U.S. Will Find Fewer Takers in the Western Pacific for Its Missiles," Stratfor, 21 August 2019, https://worldview.stratfor.com/article/us-will-find-few-takers-western-pacific-its-missiles-china-russia.

604. Jeffrey W. Hornung, "Is Japan's Interest in Strike Capabilities a Good Idea?" *War on the Rocks*, 17 July 2020, https://warontherocks.com/2020/07/is-japans-interest-in-strike-capabilities-a-good-idea/; "Australia partners with US to develop hypersonic missiles," *CNN*, 1 December 2020, https://www.cnn.com/2020/12/01/australia/hypersonic-missile-australia-us-intl-hnk/index.html.

lies may also provide opportunities to strengthen allied resolve, as heavy-handed intimidation and coercion provokes allies into taking stronger stances against American adversaries.[605] By investing in its alliances and taking advantage of rival missteps, the United States can create the political conditions under which it can exploit its advantages in theater-range weapons.

When it comes to arms limitation, the American position on theater-range weapons is ambiguous. On the one hand, the United States' advantages in theater conventional forces suggest that it should forgo arms limitation in this category and instead deploy such weapons and encourage allies to do the same.[606] On the other hand, the United States may see greater advantage in controlling theater conventional weapons, so long as those controls are equal. The cost and complexity of theater-range missiles make it difficult to amass them in sufficient numbers to coerce adversaries directly in wartime. Instead, theater conventional weapons will disrupt enemy wartime activities and create windows of opportunity for massive shorter-range forces to exploit. For example, China plans to use theater conventional missiles as part of its anti-access strategy, delaying American entry into the Western Pacific long enough for Chinese forces to invade or coerce their neighbors.[607] Similarly, the United States plans to

---

605. Jung H. Pak, "Trying to Loosen the Linchpin: China's Approach to South Korea," Brookings Global China, July 2020, https://www.brookings.edu/wp-content/uploads/2020/07/FP_20200606_china_south_korea_pak_v2.pdf; Mihir Sharma, "China's Bullying Tactics Will Only Unite Its Foes," *Bloomberg*, 19 November 2020, https://www.bloomberg.com/opinion/articles/2020-11-19/china-s-trade-pressure-on-australia-will-likely-fail; Luke Patey, "China is an Economic Bully – and Weaker Than It Looks," *Foreign Policy*, 4 January 2021, https://foreignpolicy.com/2021/01/04/china-is-an-economic-bully-and-weaker-than-it-looks/.
606. Andrew S. Erickson, "Good Riddance to the INF Treaty: Washington Shouldn't Tie Its Own Hands in Asia," *Foreign Affairs*, 20 August 2019, https://www.foreignaffairs.com/articles/china/2019-08-29/good-riddance-inf-treaty.
607. Andrew F. Krepinevich and Barry Watts, *Meeting the Anti-Access and Area-Denial Challenge* (Washington, DC: Center for Strategic and Budgetary Assessments, 2003), https://csbaonline.org/research/publications/a2ad-anti-access-area-denial; Eric Heginbotham *et al*, *Chinese Attacks on U.S. Air Bases in Asia: An Assessment of Relative Capabilities, 1996-2017* (Santa Monica, CA:

use theater conventional fires to suppress enemy anti-access systems thus allowing American ships and tactical aircraft to close with and destroy the adversary.[608] Yet the United States and its allies might be able to achieve similar results using cheaper, shorter-range weapons.[609] If theater conventional weapons were strictly limited or even eliminated, then the United States and its allies might actually be *even better off* in their ability to maneuver around China and Russia's maritime peripheries while denying that same maneuver to the Eurasian challengers.

In the realm of theater conventional weapons, the United States' geopolitical position offers it a wealth of advantages, whether it chooses to push for low and equal limitations or simply to compete vigorously with its rivals. To best exploit these advantages, however, American leaders should decide *which* approach they wish to pursue, and soon. The choice is more than academic, as it bears directly on the question of whether the United States should proliferate theater conventional weapons to its partners, or not. Proliferating missiles to allies will entail fewer alliance political problems in the shorter term, allowing a larger allied missile force to be deployed more quickly in China and Russia's vicinities. If the longer-term objective is an equal and low limitation of theater conventional weapons, however, American control would dramatically simplify

---

The RAND Corporation, 2015), https://www.rand.org/pubs/research_briefs/RB9858z2.html.

608. Jan van Tol *et al*, *AirSea Battle: A Point-of-Departure Operational Concept* (Washington, DC: Center for Strategic and Budgetary Assessments, 2010), https://csbaonline.org/research/publications/airsea-battle-concept; Sydney J. Freedberg, Jr., "Army Rebuilds Artillery Arm For Large-Scale War," *Breaking Defense*, 27 April 2020, https://breakingdefense.com/2020/04/army-rebuilds-artillery-arm-for-large-scale-war/; Douglas Birkley, "The DoD needs to rethink long-range strike in its joint war fighting concept," *Defense News*, 21 August 2020, https://www.defensenews.com/opinion/commentary/2020/08/21/the-dod-needs-to-rethink-long-range-strike-in-its-joint-war-fighting-concept/.

609. Beckley, "The Emerging Military Balance in East Asia," https://www.mitpressjournals.org/doi/pdf/10.1162/ISEC_a_00294; Eric Heginbotham and Richard J. Samuels, "Active Denial: Redesigning Japan's Response to China's Military Challenge," *International Security* 42, 4 (2018), 128-169, http://dx.doi.org/10.1162/isec_a_00313.

the ultimate deal with China and/or Russia because fewer countries would sit at the negotiating table.

Over time, range of conventional weapons may increase to intercontinental distances greater than 5,500 km.[610] While the United States *could* compete in this new category of weapon, intercontinental conventional weapons would reduce some of the utility of American allies as missile bases, since rival intercontinental missiles would be just as capable of hitting the United States. American rivals might also find such limitation attractive since conventional intercontinental missiles would compete with other acquisition priorities for scarce resources like rocket boosters and guidance systems. An approach to limit a non-existent weapon system that no power sees as militarily decisive might be a useful short-term measure to begin a multilateral arms control dialogue, even if racing continues in other missile capabilities.

Theater Nuclear Weapons: When it comes to theater nuclear weapons, the United States is not particularly strong, nor does it need to be. American leaders opted to divest from most of the theater nuclear arsenal in the 1990s, even though China and Russia did not.[611] American nuclear divestment was militarily sound, as the rapid proliferation of precision conventional weapons has rendered most theater nuclear weapons militarily redundant, even for

---

610. Bruce M. Sugden, "Speed Kills: Analyzing the Deployment of Conventional Ballistic Missiles," *International Security* 34, 1 (2009, 113-146, https://doi.org/10.1162/isec.2009.34.1.113; Amy F. Woolf, *Conventional Prompt Global Strike and Long-Range Ballistic Missiles: Background and Issues* (Washington, DC: Congressional Research Service, 2020), https://crsreports.congress.gov/product/pdf/R/R41464; Henry Sokolski, "What Missile-driven Competition with China Will Look Like," *American Purpose*, 21 October 2020, https://www.americanpurpose.com/articles/missile-driven-competition-china-sokolski/.
611. Susan J. Koch, *The Presidential Nuclear Initiatives of 1991-1992* (Washington, DC: National Defense University Press, 2012), https://ndupress.ndu.edu/Portals/68/Documents/casestudies/CSWMD_CaseStudy-5.pdf; Hans M. Kristensen, "US Navy Instruction Confirms Retirement of Nuclear Tomahawk Missile," *Federation of American Scientists*, 18 March 2013, https://fas.org/blogs/security/2013/03/tomahawk/.

America's great power rivals.[612] Despite this, theater nuclear weapons retain some *political* utility. First, countries that cannot procure enough advanced conventional weapons to meet their potential wartime needs - like Pakistan - may still rely on the battlefield effects of nuclear weapons against more powerful conventional adversaries.[613] Second, countries that fear defeat on the battlefield by conventionally-superior adversaries may still see theater nuclear weapons as useful signaling tools to terminate conflicts through escalation, regardless of the weapon's actual battlefield impact.[614] Third, countries that border great power rivals in Eurasia may use cheaper theater-range nuclear weapons to supplement their core strategic forces for mis-

---

612. Robert Tomes, "The Cold War Offset Strategy: Assault Breaker and the Beginning of the RSTA Revolution," *War on the Rocks*, 20 November 2014, https://warontherocks.com/2014/11/the-cold-war-offset-strategy-assault-breaker-and-the-beginning-of-the-rsta-revolution/; David E. Johnson, *Shared Problems: The Lessons of AirLand Battle and the 31 Initiatives for Multi-Domain Battle* (Santa Monica, CA: The RAND Corporation, 2018), https://www.rand.org/pubs/perspectives/PE301.html; Cynthia Roberts, "Revelations About Russia's Nuclear Deterrence Policy," *War on the Rocks*, 19 June 2020, https://warontherocks.com/2020/06/revelations-about-russias-nuclear-deterrence-policy/; Fiona S. Cunningham and M. Taylor Fravel, "Assuring Assured Retaliation: China's Nuclear Posture and U.S.-China Strategic Stability," *International Security* 40, 2 (2015), 7-50, https://doi.org/10.1162/ISEC_a_00215; Caitlin Talmadge, "Would China Go Nuclear? Assessing the Risk of Chinese Nuclear Escalation in a Conventional War with the United States," *International Security* 41, 4 (2017), 50-92, https://doi.org/10.1162/ISEC_a_00274; Walter C. Ladwig, III, "A Cold Start for Hot Wars? The Indian Army's New Limited War Doctrine," *International Security* 32, 3 (2007/2008), 158-190, https://doi.org/10.1162/isec.2008.32.3.158.
613. Sadia Tasleem, "Pakistan's Nuclear Use Doctrine," Carnegie Endowment for International Peace, 30 June 2016, https://carnegieendowment.org/2016/06/30/pakistan-s-nuclear-use-doctrine-pub-63913.
614. Keir A. Lieber and Daryl G. Press, "Coercive Nuclear Campaigns in the 21st Century: Understanding Adversary Incentives and Options for Nuclear Escalation," Project on Advanced Systems and Concepts for Countering Weapons of Mass Destruction, Report Number 2013-001, March 2013, https://calhoun.nps.edu/bitstream/handle/10945/34337/nps08-040813-01.pdf?sequence=1&isAllowed=y; Henry Sokolski, "A Peek into Our Nuclear Future," *The American Interest*, 5 August 2020, https://www.the-american-interest.com/2020/08/05/a-peek-into-our-nuclear-future/.

sions of coercion and counterforce.[615] To ensure robust deterrence against these sorts of provocations, the United States should retain some limited theater nuclear capabilities.[616] But there is no need to compete quantitatively with adversaries in deploying theater nuclear weapons. Nor is there any political appetite in the United States for dramatically expanding the theater nuclear footprint – new American theater-range missiles will almost certainly be conventionally-armed.[617]

Since the United States is weaker in this field, it should consider limiting theater nuclear capabilities – on its adversaries' terms. Russia especially maintains a massive legacy stockpile of tactical nuclear weapons that makes a numerically-equal agreement difficult to conceive, since none of the other great powers intend to build thousands of new theater nuclear weapons.[618] At the same time, it is difficult to imagine Russia making large unilateral cuts to its nuclear stockpile, on which Russian leaders continue to rely for both security and prestige.[619] Nor are China or India likely to accept limitations easily on weapons that they see as crucial to deterring regional rivals, includ-

---

615. Christopher P. Twomey, "Asia's Complex Strategic Environment," *Asia Policy* 11 (2011), 51-78, https://www.jstor.org/stable/24905026; Gregory D. Koblentz, *Strategic Stability in the Second Nuclear Age* (New York: Council on Foreign Relations, 2014), file:///C:/Users/John/AppData/Local/Temp/Second%20Nuclear%20Age_CSR71.pdf.
616. Keir A. Lieber and Daryl G. Press, "The Nukes We Need: Preserving the American Deterrent," *Foreign Affairs* November/December 2009, https://www.foreignaffairs.com/articles/2009-11-01/nukes-we-need; Mark B. Schneider, "Deterring Russian First Use of Low-Yield Nuclear Weapons," *Real Clear Defense*, 12 March 2018, https://www.realcleardefense.com/articles/2018/03/12/deterring_russian_first_use_of_low-yield_nuclear_weapons_113180.html.
617. Jon Harper, "Options About for New Intermediate-Range Missiles," *National Defense*, 31 October 2019, https://www.nationaldefensemagazine.org/articles/2019/10/31/options-abound-for-new-intermediate-range-missiles.
618. *Nonstrategic Nuclear Weapons* (Washington, DC: Congressional Research Service, 2020), https://fas.org/sgp/crs/nuke/RL32572.pdf#page=26.
619. Anya Loukianova Fink and Olga Oliker, "Russia's Nuclear Weapons in a Multipolar World: Guarantors of Sovereignty, Great Power Status & More," *Daedalus* 149, 2 (2020), 37-55, https://doi.org/10.1162/daed_a_01788.

ing each other.[620]

Although great powers have historically been loath to concede superiority to their rivals in arms limitation talks, in the case of theater nuclear weapons the United States would be wise to recognize its adversaries' advantages, so long as that recognition came with Chinese and Russian concessions in other fields. Specifically, the United States should offer adversaries "equivalent" theater nuclear capabilities in exchange for freezing the currently-favorable strategic nuclear balance, in which the United States has approximately as many strategic launchers as China and Russia combined.[621] Freezing this status quo – in which the United States retains its strategic nuclear advantages, while China and Russia retain their theater nuclear advantages – would ultimately play to American strengths, especially in its relations with China. Such a "freeze" of both strategic and theater nuclear capabilities might be made more enticing to China if it included not just the United States and Russia, but also regional nuclear forces like India, or latent nuclear countries like Japan. Indeed, given its paucity of theater nuclear capabilities, the United States might gain more leverage by positioning itself as a "broker" between China and its neighbors in limiting theater nuclear capabilities.

Missile Defenses and Space Forces: The United States is far ahead of its rivals in missile defense technology, and should do what it can to remain that way. As noted above, dramatic increases in the accuracy of weapons are increasing the importance of active de-

---

620. Robert Einhorn and Waheguru Pal Singh Sidhu, *The Strategic Chain: Linking Pakistan, India, China, and the United States* (Washington, DC: Brookings Institution, 2017), https://www.brookings.edu/research/the-strategic-chain-linking-pakistan-india-china-and-the-united-states/.

621. Under New START, the United States currently deploys 655 strategic launchers, while Russia deploys only 485 ("New START Treaty Aggregate Numbers of Strategic Offensive Arms," Department of State, 1 March 2020, https://www.state.gov/new-start-treaty-aggregate-numbers-of-strategic-offensive-arms-13/). China is estimated to deploy 90 intercontinental ballistic missiles and 46 submarine-launched ballistic missiles (Kristensen and Korda, "Chinese nuclear forces, 2019," https://doi.org/10.1080/00963402.2019.1628511).

fenses against missiles at all levels of conflict. The United States is moving ahead rapidly here, working to field a full spectrum of ballistic missile defenses (BMD) that can attack incoming missiles during their boost, midcourse, and terminal phases.[622] Aside from traditional BMD, concerns about the increasing lethality of conventional weapons are also driving the United States to pursue additional active defenses for bases, ships, submarines, and perhaps even aircraft.[623] Over the longer term, emerging technologies in sensing and data processing are likely to make defenses even more accurate, while new types of interceptors with deeper magazines may help alleviate cost and expand coverage.[624] Chinese and Russian missile

---

622. *Missile Defense Review* (Washington, DC: Office of the Secretary of Defense, 2019), https://www.defense.gov/Portals/1/Interactive/2018/11-2019-Missile-Defense-Review/The%202019%20MDR_Executive%20Summary.pdf; "Current U.S. Missile Defense Programs at a Glance," *Arms Control Association*, August 2019, https://www.armscontrol.org/factsheets/usmissiledefense.
623. Breaking Defense Staff, "How Active Protection Systems Knock Down Anti-Armor Threats for Both Legacy and Future Combat Vehicles," *Breaking Defense*, 17 January 2020, https://breakingdefense.com/2020/01/how-active-protection-systems-knock-down-anti-armor-threats-for-both-legacy-and-future-combat-vehicles/; Aaron Amick, "The Shadowy World of Submarine and Ship-Launched Torpedo Countermeasures," *The Drive*, 14 May 2020, https://www.thedrive.com/the-war-zone/33467/the-shadowy-world-of-submarine-and-ship-launched-torpedo-countermeasures-an-explainer; Joseph Trevithick, "Tiny Missile Interceptor to Defend Aircraft Against Enemy Missile Attacks Moves Forward," *The Drive*, 22 July 2020, https://www.thedrive.com/the-war-zone/35038/tiny-missile-interceptor-to-defend-aircraft-against-enemy-missile-attacks-moves-forward; Connie Lee, "Army Charts New Path for Air and Missile Defense," *National Defense*, 10 June 2019, https://www.nationaldefensemagazine.org/articles/2019/6/10/army-charts-new-path-for-air-and-missile-defense.
624. Jon Harper, "Just In: Pentagon Contemplating Role of AI in Missile Defense," *National Defense*, 7 October 2019, https://www.nationaldefensemagazine.org/articles/2019/10/7/pentagon-contemplating-role-of-ai-in-missile-defense; John Keller, "Military eyes prototype megawatt-class laser weapon for ballistic missile defense in next seven years," *Military & Aerospace Electronics*, 2 April 2019, https://www.militaryaerospace.com/sensors/article/16722085/military-eyes-prototype-megawattclass-laser-weapon-for-ballistic-missile-defense-in-next-seven-years; Lockheed Martin, "Next-Gen Interceptor: A New Layer of Missile Defense," *Breaking Defense*, 30 July 2020, https://breakingdefense.com/2020/07/next-gen-interceptor-a-new-layer-of-missile-defense/.

defenses are limited in comparison to those of the United States. Russia has traditionally taken a more incremental approach, seeking to add greater missile defense capabilities to its existing air defense systems.[625] Over the shorter term, China is likely to follow the Russian model of incremental modernization, though over the longer-term it may pursue more dedicated missile defenses along American lines.[626]

Because the United States leads its rivals in ballistic missile defenses, China and Russia regularly demand that the United States dismantle them.[627] Both rivals rely heavily on long-range missiles, and fear that American missile defenses could grow into an effective countermeasure.[628] Yet because the United States is *so* far ahead of its adversaries in this field, it makes little sense to compromise. Trading away such a valuable capability for concessions from the

---

625. Missile Defense Project, "Russian Air and Missile Defense," *Missile Threat*, Center for Strategic and International Studies, 14 June 2018, https://missilethreat.csis.org/system/russian-air-defense/.
626. "Chinese and Russian Missile Defense: Strategies and Capabilities," Department of Defense, 28 July 2020, https://media.defense.gov/2020/Jul/28/2002466237/-1/-1/1/CHINESE_RUSSIAN_MISSILE_DEFENSE_FACT_SHEET.PDF; Jim Garmonde, "Missile Defense Becomes Part of Great Power Competition," *DoD News*, 28 July 2020, https://www.defense.gov/Explore/News/Article/Article/2291331/missile-defense-becomes-part-of-great-power-competition/.
627. Gordon Lubold and Julian E. Barnes, "U.S. Dismisses Putin's Objection to European Missile Systems," *The Wall Street Journal*, 13 May 2016, https://www.wsj.com/articles/u-s-dismisses-putins-objection-to-european-missile-systems-1463172958; Matt Stiles, "Upset over a U.S. missile defense system, China hits South Korea where it hurts – in the wallet," *Los Angeles Times*, 28 February 2018, https://www.latimes.com/world/asia/la-fg-china-south-korea-tourism-20180228-htmlstory.html.
628. Stephen J. Cimbala and Adam Lowther, "Putin and Missile Defense Malaise: Broadening US Options," *Journal of European, Middle Eastern, & African Affairs* 2, 2 (2020), 4-20, https://media.defense.gov/2020/Jun/09/2002313275/-1/-1/1/CIMBALA.PDF; Tong Zhao, "Narrowing the U.S.-China Gap on Missile Defense: How to Forestall a Nuclear Arms Race," Carnegie-Tsinghua Center for Global Policy, 29 June 2020, https://carnegietsinghua.org/2020/06/29/narrowing-u.s.-china-gap-on-missile-defense-how-to-help-forestall-nuclear-arms-race-pub-82120.

Chinese and Russians may seem appealing, but would only be worthwhile if China or Russia offered *major* concessions in another important field, which (given their repeated demands for unilateral dismantling of American defenses) seems unlikely. Even if the United States *did* want to bargain with China and Russia over missile defenses, it would quickly run into further difficulties. American missile defenses are aimed primarily at second-tier threats like Iran and North Korea, so American leaders are likely to resist limitations on missile defense that do not also curtail the rogue state threat.[629] China and Russia will also resist limitation of their own dual-use air and missile defense systems, which both rely on to counter American airpower. The United States would do better to seek negotiating leverage in other ways, by modernizing its own strategic nuclear forces and by compromising in areas of American weakness, like theater nuclear weapons.

Another potential argument for limiting missile defenses is that such limitation would also enable controls on space weapons, since many missile defense systems also have latent anti-satellite capability.[630] If limiting missile defenses would somehow render space systems invulnerable, then it would certainly be worth pursuing, but in fact limiting missile defenses will do little to demilitarize outer space. The ongoing revolution in missile accuracy places a premium on the ability both to gather information at long ranges and to degrade the opponent's information-gathering capabilities. Missile enabling systems in outer space will therefore be increasingly important to all the great powers, as will capabilities to disrupt each other's space assets.[631] The importance of military space systems would only de-

---

629. Brad Roberts, "On the Strategic Value of Ballistic Missile Defense," *Proliferation Papers* 50 (Paris: Ifri, 2014), https://www.ifri.org/sites/default/files/atoms/files/pp50roberts.pdf.
630. Laura Grego, "The Anti-Satellite Capability of the Phased Adaptive Approach Missile Defense System," Federation of American Scientists, Winter 2011, https://fas.org/pubs/pir/2011winter/2011Winter-Anti-Satellite.pdf; Sophia Chen, "Missile-Defense Systems May Risk Space Warfare," *APS News* 27, 8 (2018), https://www.aps.org/publications/apsnews/201808/missile-defense.cfm.
631. Defense Intelligence Agency, "Challenges to Security in Space," January 2019, https://www.dia.mil/Portals/27/Documents/News/Military%20Power%20

crease over the longer term if emerging technologies decrease the military utility of outer space – for example, by replacing expensive and vulnerable satellite constellations with swarms of cheaper and easier-to-replace drones, or by improving inertial navigation systems to the point that orbital radio-navigation like GPS becomes redundant.[632] Until such time, none of the great powers are likely to surrender their space weapons, regardless of the status of missile defense programs.

Aside from the strategic appeal of anti-satellite weapons, limiting anti-satellite capabilities is especially difficult because many emerging anti-satellite technologies are dual-use. While existing kinetic anti-satellite missiles bear some similarity to American hit-to-kill missile interceptors, future anti-satellite weapons are more likely to take the form of maneuvering orbital spacecraft capable of rendezvous proximity operations (RPO), in which one satellite approaches another and interacts with it via robotic appendages.[633] RPO-capable spacecraft have important peaceful utility for repairing or enhancing existing satellites, but the same technology would also allow an attacker to disable or even destroy crucial satellites with tremendous speed and precision.[634] Similarly, adversaries might target important

---

Publications/Space_Threat_V14_020119_sm.pdf; C. Todd Lopez, "Defense Space Strategy Addresses Militarization, Competition," *DoD News*, 18 June 2020, https://www.defense.gov/Explore/News/Article/Article/2224914/defense-space-strategy-addresses-militarization-competition/; Rajeswari Pillai Rajagopalan, "From earth to space: India and China's space programs gear up for intense competition ahead," *Observer Research Foundation*, 29 June 2020, https://www.orfonline.org/research/from-earth-to-space-68717/.

632. Matt Sivacek, "Is it time? Ending the Military's Reliance on GPS," *Real Clear Defense*, 28 February 2018, https://www.realcleardefense.com/articles/2018/02/28/is_it_time_ending_the_militarys_reliance_on_gps__113131.html; Mike Cherney, "Drone Ambitions Soar to the Stratosphere," *The Wall Street Journal*, 26 August 2019, https://www.wsj.com/articles/drone-ambitions-soar-to-the-stratosphere-11566822941.

633. Brian G. Chow, "Stalkers in Space: Defeating the Threat," *Strategic Studies Quarterly* 11, 2 (2017), 82-116, https://www.airuniversity.af.edu/Portals/10/SSQ/documents/Volume-11_Issue-2/Chow.pdf.

634. Anuradha Damale, "Rendezvous Proximity Operations: Not operating in isolation," *Europna Leadership Network*, 12 August 2020, https://www.euro-

satellites with ground-based laser systems, blinding fragile sensors and perhaps eventually frying critical electrical components.[635] Like RPO-capable spacecraft, however, accurate high-powered lasers have important peaceful uses in space navigation and scientific experimentation.[636] Given their important peaceful utility, prohibition of RPO-capable spacecraft and high-powered lasers seems unlikely. The United States should ensure that it keeps abreast of these important capabilities.

Although prohibition of anti-satellite weapons seems unlikely, the United States would still benefit tremendously from pursuing clearer "rules of the road" with partners and adversaries regarding the use of anti-satellite capabilities. After all, the United States remains the world's foremost user of outer space for military missions, and so has the most to lose if space-based systems are destroyed.[637] Nor is the United States' disproportionate military reliance on outer space likely to change even as China and Russia deploy their own space-based sensing and navigation systems: after all, in the event of conflict it is the American military that will need to play the "away game" in the adversary's back yard, where adversaries can more easily rely upon ground- and air-based systems to make up for space-based losses.[638] One possible approach would be for coun-

---

peanleadershipnetwork.org/commentary/rendezvous-proximity-operations-not-operating-in-isolation/.

635. Patrick Tucker, "Russia Claims It Has Lasers to Shoot Satellites," *Defense One*, 26 February 2018, https://www.defenseone.com/technology/2018/02/russia-claims-it-now-has-lasers-shoot-satellites/146243/; Sandra Erwin, "Pentagon report: China amassing arsenal of anti-satellite weapons," *Space News*, 1 September 2020, https://spacenews.com/pentagon-report-china-amassing-arsenal-of-anti-satellite-weapons/.

636. "Satellite Laser Ranging and Earth Science," NASA Space Geodesy Program, https://ilrs.gsfc.nasa.gov/docs/slrover.pdf.

637. "UCS Satellite Database," Union of Concerned Scientists, 1 August 2020, https://www.ucsusa.org/resources/satellite-database. Over time, we might even see commercial and military exploitation of space beyond Earth's immediate orbit; see, Peter Garretson, "What War in Space Might Look Like Circa 2030-2040," Nonproliferation Policy Education Center, 13 August 2020, http://www.npolicy.org/article.php?aid=1501&tid=2.

638. Daniel Cebul, "How the Air Force plans to use space to project power in

tries to agree not to pre-position RPO-capable spacecraft near each other's satellites, perhaps by establishing a "self-defense" or "warning" zone around each satellite in which other countries would enjoy the right of "innocent passage" but within which countries would not permanently position satellites in a threatening configuration.[639] Similarly, countries could agree not to target each other's satellites with high-powered lasers in peacetime, which would at least limit the exposure of American satellites to regular adversary attacks, and agree to publicize the location of peaceful high-powered lasers to improve transparency.[640] Finally, the United States should take unilateral steps to improve the resilience of its space-based assets, especially in developing cheaper launch systems and constellations of smaller satellites better able to weather adversary attacks.[641]

Exotic Strike Technologies: Although today the great powers are scrambling to build more ballistic and cruise missiles, they are also working to develop new types of projectiles that could reshape their contest for long-range attack, conventional *and* nuclear. Much of

---

the 21st century," *DefenseNews*, 18 September 2018, https://www.defensenews.com/newsletters/digital-show-daily/2018/09/18/how-the-air-force-plans-to-use-space-to-project-power-in-the-21st-century-2/.

639. Chow, "Stalkers in Space," https://www.airuniversity.af.edu/Portals/10/SSQ/documents/Volume-11_Issue-2/Chow.pdf; Michael Cerny *et al*, "Countering Co-Orbital ASATs: Warning Zones in GEO as a Lawful Trigger for Self-Defense," Nonproliferation Policy Education Center, 30 June 2020, http://www.npolicy.org/article.php?aid=1485&tid=2. The French government has already announced self-defense zones around its critical satellites (Brian G. Chow and Henry Sokolski, "The United States should follow France's lead in space," *Space News*, 9 September 2019, https://spacenews.com/the-united-states-should-follow-frances-lead-in-space/.

640. Brian G. Chow and Henry Sokolski, "U.S. satellites increasingly vulnerable to China's ground-based lasers," *Space News*, 10 July 2020, https://spacenews.com/op-ed-u-s-satellites-increasingly-vulnerable-to-chinas-ground-based-lasers/.

641. Theresa Hitchens, "Build, Upgrade Satellites Fast for Space Resilience: Aerospace Corp.," *Breaking Defense*, 24 March 2020, https://breakingdefense.com/2020/03/build-upgrade-satellites-fast-for-space-resilience-aerospace-corp/; Dax Linville and Robert A. Bettinger, "An Argument against Satellite Resiliency: Simplicity in the Face of Modern Satellite Design," *Air & Space Power Journal* 34, 1 (2020), 43-53, https://www.airuniversity.af.edu/Portals/10/ASPJ/journals/Volume-34_Issue-1/V-Linville_Bettinger.pdf.

this contest for future capabilities is focused on building faster and more lethal missiles, especially new hypersonic weapons that combine the speed of ballistic missiles with the maneuverability and evasion of cruise missiles.[642] Since hypersonic technology is still in its infancy, many have called for preemptive arms control to prevent its emergence.[643] Yet the technology's very immaturity makes it hard to assess, since it remains to be seen which of the great powers will predominate in this field.[644] If the contest for hypersonic weapons remains close, then the United States might find opportunities to use arms limitation to enhance whatever relative American advantages emerge, but probably not until such weapons are thoroughly tested and fielded. Nor are adversaries likely to favor limitation, especially since China and Russia link their budding hypersonic programs to the need to counter American ballistic missile defenses, which the United States should not agree to limit.[645]

---

642. Michael T. Klare, "An 'Arms Race in Speed': Hypersonic Weapons and the Changing Calculus of Battle," *Arms Control Association*, June 2019, https://www.armscontrol.org/act/2019-06/features/arms-race-speed-hypersonic-weapons-changing-calculus-battle; Richard Stone, "'National pride is at stake.' Russia, China, United States race to build hypersonic weapons," *Science*, 8 January 2020, https://www.sciencemag.org/news/2020/01/national-pride-stake-russia-china-united-states-race-build-hypersonic-weapons; *Hypersonic Weapons: Background and Issues for Congress* (Washington, DC: Congressional Research Service, 2020), https://fas.org/sgp/crs/weapons/R45811.pdf.
643. John Borrie, Amy Dowler, and Pavel Podvig, *Hypersonic Weapons: A Challenge and Opportunity for Strategic Arms Control* (New York: The United Nations, 2019), https://www.un.org/disarmament/publications/more/hypersonic-weapons-a-challenge-and-opportunity-for-strategic-arms-control/; Douglas Barrie, "Unstable at speed: hypersonics and arms control," *IISS Military Balance Blog*, 18 October 2019, https://www.iiss.org/blogs/military-balance/2019/10/hypersonics-arms-control; Cameron Tracy, "Fitting Hypersonic Weapons into the Nuclear Arms Control Regime," *All Things Nuclear*, 1 April 2020, https://allthingsnuclear.org/ctracy/fitting-hypersonic-weapons-into-the-nuclear-arms-control-regime.
644. Steven Pifer, "As US-Russian arms control faces expiration, sides face tough choices," Brookings Institution, 23 March 2020, https://www.brookings.edu/blog/order-from-chaos/2020/03/23/as-us-russian-arms-control-faces-expiration-sides-face-tough-choices/.
645. Erika Solem and Karen Montague, "Updated – Chinese Hypersonic Weapons Development," *China Brief* 16, 7 (2016), https://jamestown.org/program/

While hypersonics have attracted the most attention in recent years, there is also good reason to be skeptical of their ultimate utility. Hypersonic weapons may enjoy some advantages over existing ballistic missiles, but the high-performance hypersonic weapons being tested today will remain subject to the existing economics of long-range strike, in which increases in weapon range and complexity inevitably entail decreases in volume of fire.[646] Even if hypersonic weapons have a role, therefore, their cost means that they will not replace traditional ballistic and cruise missiles, any more than those missiles replaced shorter-range gravity bombs and artillery. Instead, truly transformative long-range strike will require upending current missile economics by creating weapons that can deliver high volumes of fire at low cost over long ranges. Such transformative weapons might take the form of much cheaper variants of existing missiles, such as reusable boosters launching ballistic or hypersonic warheads.[647] But future high-volume, long-range weapons might also be entirely different from existing missiles, perhaps through combining advances in robotics and machine autonomy to create swarms of intelligent loitering weapons.[648] As with hypersonics, the rapid technical progress in autonomous weapons makes it difficult to assess which great power leads its development.[649] However, the

---

updated-chinese-hypersonic-weapons-development/; Patrick Tucker, "Pentagon Seeks a Way to Shoot Down Putin's 'Invincible' Hypersonic Missiles," *DefenseOne*, 28 January 2020, https://www.defenseone.com/technology/2020/01/pentagon-seeks-way-shoot-down-putins-invincible-hypersonic-missiles/162717/.

646. Ivan Oelrich, "Cool your jets: Some perspective on the hyping of hypersonic weapons," *Bulletin of the Atomic Scientists* 76, 1 (2020), 37-45, https://doi.org/10.1080/00963402.2019.1701283.

647. Jeff Becker, "When It Comes to Missiles, Don't Copy Russia and China – Leapfrog Them," *War on the Rocks*, 30 June 2020, https://warontherocks.com/2020/06/when-it-comes-to-missiles-dont-copy-russia-and-china-leapfrog-them/.

648. Paul Scharre and Michael C. Horowtiz, *An Introduction to Autonomy in Weapon Systems* (Washington, DC: Center for a New American Security, 2015), https://www.files.ethz.ch/isn/188865/Ethical%20Autonomy%20Working%20Paper_021015_v02.pdf; Zachary Kallenborn, "The Era of the Drone Swarm is Coming, and We Need to be Ready for It," *Modern War Institute*, 25 October 2018, https://mwi.usma.edu/era-drone-swarm-coming-need-ready/.

649. Adrian Pecotic, "Whoever Predicts the Future Will Win the AI Arms Race,"

United States certainly has the most experience operating robotic attack aircraft over large areas at long ranges, which could prove an important advantage over less experienced adversaries.[650] In the meantime, the United States should do more to assess the proliferation of advanced robotics technology, including dual-use commercial technology, to retain what advantages it can in this field.[651]

Conclusion

The competitive arms control negotiations of the coming years will be much more trying than the autopilot negotiations of the post-Cold War period. We should expect the great powers to strive for advantage in emerging military technologies, and to use negotiations as tools to shape the balance of power in their favor. If arms limitation is to have continued relevance to American national security, then we need to take steps now to reorient our arms control policy and incorporate it into a holistic strategic approach. To this end, American leaders should keep in mind the following advice.

Prepare for Rivalry: Competitive arms control thrives on a cycle of contest and negotiation, each of which feeds off of the other. Successful military-technical preparation drives arms limitation negotiations; if you don't compete, you don't get to negotiate. American leaders should take steps necessary to modernize key forces and expand into new capabilities. Negotiations will also drive future rounds of military-technical contest. American negotiators need to bargain hard and be willing to walk away from bad compromises

---

Foreign Policy, 5 March 2019, https://foreignpolicy.com/2019/03/05/whoever-predicts-the-future-correctly-will-win-the-ai-arms-race-russia-china-united-states-artificial-intelligence-defense/.

650. Larry Lewis and Diane Vavrichek, *Rethinking the Drone War: National Security, Legitimacy, and Civilian Casualties in U.S. Counterterrorism Operations* (Quantico, VA: Marine Corps University Press, 2016), https://fas.org/man/eprint/drone-war.pdf.

651. Export Controls Working Group, *UAV Export Controls and Regulatory Challenges* (Washington, DC: Stimson, 2015), https://www.stimson.org/wp-content/files/file-attachments/ECRC%20Working%20Group%20Report.pdf.

that would undermine American power. Long-term military-technical advantages must be both the first step and the final goal towards a new, sustainable framework of arms control.

The Soviet Union's historical demand for "equivalent" forces should also underline the importance of husbanding and sustaining American advantages. Although demands for "equivalent" forces are often smokescreens for self-serving proposals, Eurasian rivals are correct to worry that the United States enjoys significant technological and geopolitical advantages over them. While China and other countries have made significant inroads challenging American technical superiority, they have been far less effective at replicating America's advantageous political geography, including its vast network of alliances and security partners. Any long-term strategy for advantage must lean *into* those advantages, rather than alienating or disengaging from our alliance network.

Deploy Arms Limitation Strategically: Arms limitation negotiations offer tremendous opportunities to enhance American strengths and weaken rivals, but only if American leaders understand their relative strengths and structure negotiations accordingly. Where the United States enjoys significant, durable advantages – such as missile defenses and perhaps theater conventional weapons – it should avoid arms limitation negotiations, and pursue its strengths. Where the United States is strong – such as strategic nuclear weapons – it should pursue arms limitation on the basis of strict equality with adversaries, ideally its existing rough equality with *both* Chinese and Russian forces, but possibly eventual equality with *each*. Where the United States is not strong – such as theater nuclear weapons – it should offer to ratify adversary advantages in exchange for more important objectives, like strategic nuclear advantages. Finally, the United States should not rush into negotiations prohibiting emerging technologies until the impact of those technologies on the balance of power is better understood. American leaders should instead seek normative restrictions that shore up potential vulnerabilities, especially in the cyber and space domains.

Lay the Multilateral Groundwork: Multilateral great power arms limitation is a laudable goal, but it is also one that will only be realized over the long term and with significant American effort. Although successful negotiations are unlikely to emerge in the near future, there are still important steps the United States can take to promote its arms limitation agenda. First, the United States must insist on Chinese participation in arms control talks. China will undoubtedly resist, but the United States should continue to advance the idea that China's status as a "responsible stakeholder" requires participating in arms limitation discussions. Second, the United States should double down on relations with sympathetic partners in future negotiations, especially India. This should include expanding Indian-American dialogue to discuss each country's expectations in future arms limitation negotiations, both for notional bilateral agreements and potential multilateral agreements with third parties, with an eye towards positioning India as a partner in future talks with China. Similar discussions with traditional partners including Australia, Britain, France, Germany, Israel, Japan, and South Korea would also be wise, both to enlist their aid in talks, and to ensure that rivals cannot use negotiations as a ploy to divide the United States from its partners.

Third, the United States should not neglect Russian-American bilateral dialogue. Although the Cold War negotiating framework is crumbling in the face of new geopolitical realities, properly-structured Russian-American discussions can have continuing relevance in the new era of multilateral rivalry. Despite their differences, the United States and Russia both benefit from competing less with each other and focusing more on China. Extending New START is a good first step to ensure that the United States can direct resources towards long-term strategic rivalry with China, rather than racing Russia in the short term to retain aging legacy systems.

About the Authors

David Bernstein is from Salt Lake City, Utah and will graduate from Georgetown University's Walsh School of Foreign Service (Class of 2021), majoring in International Politics with a concentration in International Security. His research interests and his upcoming thesis concentrate on nonproliferation efforts towards allies under the US "nuclear umbrella." Along with his coursework, David works as a research assistant for Dr. Caitlin Talmadge, Dr. Brandon Valeriano, and Dr. Eric Borghard, focusing on nuclear deterrence and cyber strategy. David has interned at the Nuclear Threat Initiative, in a congressional office, and on multiple campaigns. As a member of the Georgetown policy debate team, David won 2nd place at the 2017 National Debate Tournament college championship and has been ranked in the top 16 partnerships in the country.

Michael B. Cerny is currently pursuing his Master of Philosophy in Politics (Comparative Government) at the University of Oxford, where he is specializing in the politics and governance of China. He earned a bachelor's degree in Political Science with a double major in Chinese Studies from Emory University, where he graduated Phi Beta Kappa with the Emory College Language Center's Excellence in Language Studies Award (Chinese). At Emory, Cerny co-founded the *Emory Journal of Asian Studies* and oversaw the publication of its first two issues. Cerny is an active volunteer for The Carter Center's initiatives related to US-China relations.

Brian Chow is an independent policy analyst and author of over 160 publications. He is a former Senior Physical Scientist at the RAND Corporation where he worked for 25 years. He was appointed by the President's Science Advisor, the Under Secretary of Defense for Policy, and the Office of the Chief of Naval Operations as a consultant to advise them on policy matters. His recent articles on space appear in *Strategic Studies Quarterly, SpaceNews, The National Interest, Defense One, Defense News*, and *The Space Review*. He holds a Ph.D. in physics from Case Western Reserve University and an MBA (with Distinction) and Ph.D. in finance from the University of Michigan. He can be reached at brianchow.sp@gmail.com and on Twitter at @briangchow.

David A. Cooper holds a tenured faculty appointment at the US Naval War College as The James V. Forrestal Professor of National Security Affairs and previously served as Chair of the National Security Affairs Department. His areas of expertise include arms control and disarmament, nonproliferation, multilateral negotiations and organizations, international relations theory, and foreign policy analysis. He has published various articles, chapters, and books in these areas, including most recently this year as co- author of the graduate textbook *Decision-Making in American Foreign Policy: Translating Theory into Practice* from Cambridge University Press. He has a forthcoming book from Georgetown University Press on *Arms Control for the Third Nuclear Age: Between Disarmament and Armageddon*. He holds a PhD in Political Science and International Relations from The Australian National University (ANU), a Master of International Affairs from Columbia University, and is a graduate of Oberlin College. Dr. Cooper is a scholar-practitioner. Before becoming an academic he served for nearly two decades as an official in the Office of the Secretary of Defense (OSD) where he was a career member of the Senior Executive Service (SES). He held several key positions including as Director of Nonproliferation Policy and Director of Strategic Arms Control Policy. He first came to the Pentagon in 1990 as a Presidential Management Intern.

Peter Garretson is an independent strategy consultant who focuses on space and defense, and a senior fellow in Defense Studies with the American Foreign Policy Council. He was previously the director of Air University's Space Horizons Task Force, America's think tank for space, and was deputy director of America's premier space strategy program, the Schriever Scholars. He is the author of a forthcoming book on great power competition for space resources.

Brandon Kelley is a graduate student in Georgetown University's Security Studies Program, completing his final year with a concentration in Military Operations. Brandon's work focuses on strategic deterrence issues, especially in the space and counterspace contexts. He began conducting independent open-source research and analysis of counterspace capabilities while pursuing a B.A. in Science, Tech. & Int'l Affairs at Georgetown, with a focus on the threat posed by co-orbital ASATs. After graduation, he played a key role in research and writing for the inaugural edition of the Secure World Foundation's *Global Counterspace Capabilities: An Open Source Assessment*, now one of the primary resources for publicly-available information on the subject. Since 2017, Brandon has served full-time as the Associate Director of Debate at Georgetown, where he coaches and directs strategy, argument development, and research for the multiple-time national championship-winning Policy Debate Team. After four years of working to secure a strong financial and competitive future for the program, Brandon looks forward to shifting to full-time involvement in space security issues in the coming year.

John D. Maurer is a professor at the School of Advanced Air and Space Studies (SAASS) and a visiting scholar at the American Enterprise Institute (AEI). His research focuses on nuclear and grand strategy, arms control, national security, and US diplomatic history. Dr. Maurer's work has been published by a variety of academic journals and media outlets, including *Diplomatic History, Texas National Security Review, The National Interest*, and *War on the Rocks*. In his forthcoming book, *Competitive Arms Control: Nixon, Kissinger, and SALT, 1969-1972*, he examines the strategy behind the Strategic Arms Limitation Talks (SALT I) in the Richard Nixon administration. Dr. Maurer has a BS in international politics and a PhD in US diplomatic history from Georgetown University.

Xavier Pasco has been Director of the Foundation for Strategic Research (FRS) since October 2016. Dr. Pasco has coordinated research on Space, High Technology and Security at FRS and at CREST (Center for Research and Evaluation of the relationships between Strategies and Technology) associated to Ecole Polytechnique. He is a specialist of the United States space policy and programs, with a focus on the transatlantic relationships involving military space and interoperability issues. He has also been in charge of studies on future national and European programs in the space sector, both in the civil and defense fields. In this capacity, he is the author of numerous reports on civil and military space activities carried out on behalf of national and European public bodies. Xavier Pasco has been a member of multiple ministerial and inter-ministerial working groups. Dr. Pasco participated in several projects related to the "Preparatory Action for Security Research" launched by the European Commission in 2004, and within the framework of the 6th and 7th European Research and Development Framework Programme (2007-2013). In this context, he coordinated the European cross-disciplinary research activity "Conditions for Space Policy and Related Action Plan Consolidation in Europe" supported by the Executive Research Agency of the European Union in 2010-2011. He also coordinated European work on data policy for the future European space surveillance system as part of a study for the European Space Agency (ESA). He also coordinated (2011-2012) the study on the future European ground segment for earth observation for defense and security purposes for the European Defence Agency (EDA). He is today the lead coordinator of SPACEWAYS, a European Commission-sponsored project dedicated to studying Space Traffic Management and started in January 2021 with a team of key European industry actors, research centers and Institutes. Before becoming FRS Director, Xavier Pasco has been a Senior Researcher at the Foundation for Strategic Research (since 1994). He was a lecturer at the Institut d'Etudes politiques de Paris and an Associate Professor in Political Science at the University of Marne-la-Vallée. He was also a Research Fellow at CREST-Ecole Polytechnique between 1989 and 1994. Xavier Pasco has been an Associate Research Fellow at the Space Policy Institute at George Washington University (Washington D.C., USA) since 1994. He is the European Editor-in-Chief of the international quarterly journal "Space Policy" (Elsevier Science). He is an expert at the European Economic and Social Council and a member of the International Academy of Astronautics. He is also a member of the US based Secure World Foundation (SWF) advisory committee. Xavier Pasco holds a PhD in political science from the University of Paris I Panthéon-Sorbonne.

Raphael J. Piliero is a graduate of Georgetown University, with a Bachelor of Arts in government. He currently works with the Forward Defense practice at the Atlantic Council's Scowcroft Center for Strategy and Security through the Young Global Professionals program. In this role, Piliero specializes in the practice's nuclear and space portfolios and was a named contributor to the Atlantic Council's first space strategy paper. During his time at Georgetown, he assisted with research on Chinese strategic forces for Professor Matthew Kroenig while also interning with the House Foreign Affairs chairman and the United Nations Association. Piliero's writings have been featured in *The Yale Review of International Studies, The Diplomat, Defense News, The US-China Perception Monitor*, and *Political Wire.*

Henry D. Sokolski is executive director of the Nonproliferation Policy Education Center and teaches graduate classes on nuclear policy at the University of Utah and the Institute of World Politics. He was recently appointed Senior Fellow for Nuclear Security Studies at the University of California at San Diego. Previously, he worked in the Pentagon as Deputy for Nonproliferation Policy, as a consultant to the National Intelligence Council, as a member of the Central Intelligence Agency's Senior Advisory Group, as a Senate military legislative aide to a member of the Senate Armed Services Committee, and as a special assistant on nuclear energy to the chairman of Senate's Tennessee Valley Authority oversight committee. He served on two congressional commissions on the prevention of WMD proliferation and has authored and edited numerous volumes on strategic weapons proliferation issues, including *Best of Intentions: America's Campaign against Strategic Weapons Proliferation* and *Underestimated: Our Not So Peaceful Nuclear Future.* Mr. Sokolski earned his graduate degree in political science from the University of Chicago.

Made in the USA
Monee, IL
04 November 2026

67386529R00157